Tarascan Copper Metallurgy:

A multiapproach perspective

Blanca Estela Maldonado

Archaeopress Publishing Ltd
Summertown Pavilion
18-24 Middle Way
Summertown
Oxford OX2 7LG

www.archaeopress.com

ISBN 978 1 78491 625 1
ISBN 978 1 78491 626 8 (e-Pdf)

© Archaeopress and Blanca Estela Maldonado 2018

All rights reserved. No part of this book may be reproduced, or transmitted, in any form or by any means, electronic, mechanical, photocopying or otherwise, without the prior written permission of the copyright owners.

Printed in England by Oxuniprint, Oxford

This book is available direct from Archaeopress or from our website www.archaeopress.com

Contents

List of Figures and Tables iii

Preface v
    References cited vi

Chapter 1 Introduction 1
    Problem definition and research design issues 1
        Methodology 1
        Data collection and sampling 2
        Significance 2
    The approach: metallurgy as technology 3
        Specific issues covered in this volume 4
    Chapter organization and content 5

Chapter 2 Approaches to the study of technology and craft production 7
    The definition of technology in archaeology: historical background 7
    Theoretical approaches to technology and culture 8
        The historical/particularist approach 8
        The evolutionary approach 9
        Neo-Darwinian approaches 9
        Behavioral and economic factors 10
    Methodological approaches to technology 11
        Models derived from evolutionary theory 11
            Applicability of evolutionary models 12
            Implementing the Theory: The Study of Technological Processes 13
        Chaîne opératoire: the concept and its application 13
    Perspectives of the present study: technology, craft production and political economy 15
        Approaches to the study of craft production in archaeology 16
        The Organization of craft production: political ecology vs. political economy 16
    Discussion 17

Chapter 3 Synopsis of preindustrial metallurgy as applied to Mesoamerica 19
    Background: the emergence of metallurgy in the New World 19
    Mining, metallurgy, and the evidence for West Mexico 20
        Native metals and ore minerals in West Mexico 21
        Prospecting and mining 22
        Extractive metallurgy 25
            Beneficiation 25
            Pyrometallurgy 26
            Alloy technology 28
        Fabrication of artifacts 29
            Cold and hot working 30
            Lost-wax casting 34
    Discussion 36

Chapter 4 Tarascan copper smelting at the zone of Itziparátzico: a case study 38
    Socio-technological background: the Tarascan state 38
        Metal and the political economy of the Tarascan state 38
        Patterns of distribution and consumption of metal 40
        Ore deposits and mining in the Tarascan territory 40
        Smelting activities in the Tarascan territory 41
    Physiographic background: the Santa Clara-Opopeo Region 43
        Hydrology and climate 44
        Soils and vegetation 44
        History of land use 45

    Itziparátzico: the research project 46
        Preliminary work 48
        Surface survey 48
        Test pitting 51
        Summary description of Itziparátzico 55
        Slag analysis 55
            Sample selection, preparation and processing 56
            Microscopic analyses of slag samples 57
            Summary results of the microscopic analyses 57
            Chemical composition analysis of the slag samples 58
            Summary of the results of the chemical composition analysis 59
        Interpretation of the scientific analyses of the slag samples 60
    Discussion 61

Chapter 5 Models of technological organization 63
    The study of craft production 63
        Some notes on attached vs. independent craft production 64
            Gradients of variation in Costin's Scheme 64
        Scale of production units 66
        What was the scale of production at Itziparátzico? 67
    Organization of metallurgical production: two models 68
        Organizational Model I: local metallurgical industry 68
            Archaeological correlates 68
            An ethnohistorical example from south-central Africa 69
        Organizational Model II: mobilized local metallurgical industry 69
            Archaeological correlates 70
            The Kingdom of Benin: an ethnohistorical example from southwestern Nigeria 70
    The organization of copper production in the Tarascan state: local versus mobilized industry 70
        Ore deposits and mining 71
        Smelting operations 71
            Independent or supervised smeltings? 72
        Final processing and manufacturing activities 73
    Discussion 73

Chapter 6 Conclusions, remarks and suggestions for future research 76
    Evaluating the data from Itziparátzico 76
        Theory of technology 77
        Archaeometallurgy 77
        Archaeometallurgical materials 77
        The organization of copper metallurgy 78
    Potential for future research in and around Itziparátzico 78
        Technological and chronological implications 78
        Comparative models for the use of wind power in preindustrial smelting 79
            A brief overview of wind-powered smelting technology 79
        Other pressing research needs: fuel production studies and ethnohistory 80
    Potential for future research at the regional level 80
        Colonial studies as a possible avenue to investigate Prehispanic technology 81
    Conclusions 82

Bibliography 84

Appendix A IARP 2003-4: Survey data 99

Appendix B IARP 2003-4: Test pitting data 103

Appendix C Slag analyses 112

Appendix D The rise of the theory of technology 140
    V. Gordon Childe: Technology and Social Evolution 140
    Problems in Childe's Theory of Technology 142

List of Figures and Tables

Figure 1.1 Location of Itziparátzico within the Zirahuén Basin and the state of Michoacán 2
Figure 2.1a The *chaîne opératoire* approach provides a framework capable of linking material and social practices 14
Figure 2.1b The *chaîne opératoire* includes feedback loops in that the intended use of tools will affect the choice of technology and raw material. 14
Figure 3.1 West Mexico in the context of Mesoamerica. 21
Figure 3.2 Medieval mining techniques showing: three vertical shafts of which the first, A, does not reach the tunnel; the second, B, reaches the tunnel; to the third, C, the tunnel has not yet been driven. D-Tunnel 25
Figure 3.3 Experimental smelting at Cerro Huaringa, Peru, based on archaeological data from Batán Grande 27
Figure 3.4 Mesoamerican crucible and blow-pipes 28
Figure 3.5 Mesoamerican crucible and blow-pipe. 28
Figure 3.6 Mesoamerican crucible and blow-pipe, from the Codex Mendoza. 28
Figure 3.7 Replica of a Tarascan spiral tweezer. 31
Figure 3.8 Tarascan priest wearing a metal tweezer. 32
Figure 3.9 Examples of experimental axes cast in copper and tin bronze. 33
Figure 3.10 Casting of prehispanic axes, ... 34
Figure 3.11a Examples of experimental wirework bells cast in copper and tin bronze. 35
Figure 3.11b Details of experimental wirework bells cast in copper and tin bronze. 35
Figure 3.12 Flow chart describing the process of casting experimental copper and bronze bells. 36
Figure 3.13 Cycle of copper production and working 37
Figure 4.1 The Tarascan Empire within the framework of other Mesoamerican societies 39
Figure 4.2 Maximum extension of the Tarascan territory in 1522 39
Figure 4.3 Established chronology for the Pátzcuaro Basin 40
Figure 4.4 Mining centers in the Central Balsas Basin 41
Figure 4.5a *Cendrada* lined with oak ashes, used by contemporary coppersmiths from Santa Clara del Cobre 42
Figure 4.5b Coppersmith from Santa Clara del Cobre melting scrap copper. 42
Figure 4.5c Resulting red-hot copper *tejo* (ingot). 42
Figure 4.6 Santa Clara del Cobre-Opopeo 43
Figure 4.7a View of Itziparátzico 47
Figure 4.7b View of Itziparátzico from the terraces. 47
Figure 4.8 Location of Itziparátzico and other mining and smelting localities within the Tarascan territory 49
Figure 4.9 Three main sectors of the Itziparátzico area 50
Figure 4.10 Lithics from Itzparátzico: Gray-black obsidian arrow head, modified flake and prismatic blade. 52
Figure 4.11 Lithics from Itziparátzico: Gray-black obsidian core. 52
Figure 4.12 Lithics from Itziparátzico: Basalt tools. 52
Figure 4.13 Lithics from Itziparátzico: Basalt axe. 52
Figure 4.14 Polished red slip ware from Itziparátzico. 53
Figure 4.15 Red slipped stirrup spouted vessel fragment. 53
Figure 4.16 Red-and-White-on-Cream rim and body fragment. 53
Figure 4.17 Red-and-White-on-Cream body fragment. 53
Figure 4.18 Red ware with resist (negative) decoration. 53
Figure 4.19 Decorated Red-and-Black-on-White pottery. 53
Figure 4.20 Decorated Black on White pottery. 54
Figure 4.21 Decorated White on Red pottery. 54
Figure 4.22 Potsherd showing incised decoration. 54
Figure 4.23 Potsherd showing appliqué decoration. 54
Figure 4.24 Zoomorphic pipe fragment. 54
Figure 4.25 Pipe fragment showing incised decoration. 54
Figure 4.26 Lumpy slag fragment from Itziparátzico. 56
Figure 4.27 Platy slag fragment from Itziparátzico. 56
Figure 4.28 Low magnification images of a typical 'lumpy' slag sample. 57
Figure 4.29 Identification of the copper ore: chalcopyrite. 58
Figure 4.30 Scanning electron microscope (SEM) images of a 'typical' region of a lumpy slag. 58
Figure 4.31 XRF analysis of copper slag from Itziparátzico. 59
Figure 4.32 Triangular diagram showing phase relationships in the system iron oxide- Al_2O_3- SiO_2 60
Figure 5.1 Typologies of the organization of craft production 67
Map A.2 Location of mounds in the surveyed area 100
Map B.1 Location of the test pits at Itziparátzico 104
Table A.1 Recovered materials: survey 102
Table B.1 Materials recovered from excavation 108
Table B.2 Ceramic inventory 110

Preface

The present volume is based on my PhD dissertation, which I completed at The Pennsylvania State University in 2006. It outlines the design, implementation and results of an archaeological project carried out at Itziparátzico, a Tarascan locality near Santa Clara del Cobre, Mexico, where evidence indicates that copper metal production took place from the Late Postclassic (AD 1350-1520) throughout the Contact period, and continues until today.

At the time this research project began, there seemed to have been very few attempts to carry out systematic studies on primary metallurgical production at Mesoamerican sites. These include the work of Dorothy Hosler (2009) at El Manchón and other sites in Guerrero, which was only partially published several years later. The results presented in this work, for the first time characterize smelting byproducts (slag). Based on multiple sources of data, including archaeological, ethnohistorical, and archaeometrical, a model for copper metallurgical production at the local and regional level is proposed. Although there has been a considerable increase in research on Mesoamerican metallurgy during the past twelve years, the overall picture of metallurgical processes has not changed considerably.

After the conclusion of the thesis, further analyses of the results were produced, and derived in a journal article on the scale of copper production at Itziparátzico (see Maldonado and Rehren 2009). The work concludes that copper smelting at the site was done by part-time specialists embedded in a predominantly agricultural economy, and formed part of a centrally organized network of mining, smelting and processing of copper to supply the Tarascan state.

Perhaps one of the most exciting research developments to date involves new studies of smelting slags from seven archaeological sites in the surroundings of Santa Clara del Cobre, which have allowed us to, for the first time in the Mesoamerican region, obtain dates of metallurgical materials by archaeomagnetic means. The results obtained thus far suggest a continuity of metallurgical production in this area, from prehispanic times to the Colonial period. The dates obtained for Itziparátzico go as far back as 100 years before the Spanish conquest (Morales et al. 2016; Punzo et al. 2015). While not directly related to metallurgical processes, these data have allowed us to confirm the chronological framework of the site.

In terms of chronology, it is also important to address recent developments related to the dates established by Hosler (1994) for the early spread of metallurgy in Western Mexico. While her estimated time frame for copper production is A.D. 600-800, new assessments based on regional data suggest an initial date around A.D. 875 (see Nelson et al. 2015: 43).

Another important contribution to our understanding of primary smelting is provided by Aaron Shugar and colleagues (see Urban et al. 2013), who report the presence of a copper-processing area at the site of El Coyote, northwestern Honduras (A.D. 800-1000). The authors recorded a number of features that they argued represent evidence of a workshop, including slag fragments, and a presumed copper-smelting furnace. The available data are inconclusive, particularly with respect to chronology, as certain components of the workshop may postdate the early sixteenth century. Regardless, the data produced would be equally relevant to the study of preindustrial metallurgy.

Further attempts to contribute to an hypothetical reconstruction of the *chaîne opératoire* for copper smelting are currently being made in collaboration with one of my Master's students, Patricia Castro. This ongoing research involves smelting experiments conducted in purpose-built workshops, modeled from archaeological and ethnoarchaeological data. We seek to establish a technological link between the method utilized by local artisans today for melting scrap copper and the archaeological deposits near Santa Clara del Cobre. This method involves the use of shallow holes (known locally as *cendradas*) covered with ash, where alternate layers of charcoal and pieces of copper are deposited. This type of 'furnace' is mentioned in Colonial documentary sources as a native method for copper smelting. Some of the goals of the experimental program include the smelting of copper ore using the *cendrada* and the production of slag, followed by comparison of the microstructural, mineralogical, and chemical characteristics of the archaeometallurgical and experimental residues. The main objective of the research is the development of a model for the Late Postclassic Tarascan process of copper production.

In conclusion, evidence for Mesoamerican extractive copper Mesoamerican metallurgy is as elusive as it was twelve years ago. Further archaeological work is needed at Itziparátzico and other sites in the region, as well as in the mining areas. A comprehensive regional investigation of metallurgical technology would allow further testing of some of the premises advanced in the study presented here. Plans for further research are in progress.

References cited

Hosler, Dorothy   2009 West Mexican Metallurgy: Revisited and Revised. *Journal of World Prehistory* 22: 185-212.

Maldonado, Blanca and Thilo Rehren 2009 Early copper smelting at Itziparátzico, Mexico. *Journal of Archaeological Science* 36 (9): 1998–2006.

Morales, Juan, María del Sol Hernández-Bernal, Avto Goguitchaichvili, and José Luis Punzo-Díaz 2017 An Integrated Magnetic, Geochemical and Archeointensity Investigation of Casting Debris from Ancient Metallurgical Sites of Michoacán, Western Mesoamerica. *Studia Geophysica et Geodaetica* 61: 1-20.

Nelson, Ben A., Elisa Villalpando Canchola, José Luis Punzo Díaz, and Paul E. Minnis 2015 Prehispanic Northwest and Adjacent West Mexico, 1200 B.C.–A.D. 1400: An Inter-Regional Perspective. *Kiva* (81): 1-2, 31-61.

Punzo Díaz, José Luis, Juan Morales and Avto Goguitchaichvili 2015 Evidencia de Escorias de Cobre Prehispánicas en el Área de Santa Clara del Cobre, Michoacán, Occidente de Mexico. *Arqueología Iberoamericana* 28: 46-51.

Urban, Patricia, Aaron N. Shugar, Laura Richardson, and Edward Schortman 2013 The Production of Copper at El Coyote, Honduras Processing, Dating, and Political Economy. In *Archaeometallurgy in Mesoamerica: Current Approaches and New Perspectives*, edited by Scott E. Simmons y Aaron N. Shugar: 77-112. Boulder: The University of Colorado Press.

Chapter 1

Introduction

Metals have played a significant role in society throughout the ages, their use often affecting the course of civilization and thus human history. However, because 'technologies and their products shape and are shaped by economic and political interests, social values, and other elements of culture' (Hosler 1994: 3), the form and degree of this impact can be highly variable. By the sixteenth century, Old World metallurgy had evolved into a science whose development foreshadowed the Industrial Revolution. In the New World, metallurgical technologies were less fundamental but they had established their important role in the political economy of Pre-Columbian Polities. Eurasia is the area in which technology started its post-Pleistocene acceleration and resulted in the greatest accumulation of inventions (Diamond 1999). Mesoamerica, in contrast, lacked domesticated large mammals and native societies had to rely on relatively simple technologies. For this reason, they are considered low-energy societies (Webster et al. 1993). Nevertheless, although limited by a relatively simple technology, prehispanic metalworkers developed a sophisticated non-ferrous metallurgy.

Mesoamerican copper metallurgy developed in West Mexico sometime between AD 600 and 800, and over the next 900 years a wide variety of artifacts was produced. At the time of the Spanish Conquest the main locus of metal production in Mesoamerica was the Tarascan region of western Mexico. Scholars have argued that mining and metallurgy evolved into a state industry, as metal adornments used as insignias of social status and public ritual became closely associated with political control. In spite of its importance, however, Tarascan metallurgy is poorly documented. The extractive processes involved and the organization of the different aspects of this production are virtually unknown. The present work outlines the design, implementation and results of an archaeological project carried out at Itziparátzico, a Tarascan locality near Santa Clara del Cobre, Mexico, where evidence indicates that copper metal production took place from the Late Postclassic throughout the Contact period, and continues until today.

This pioneer research has required the employment of multiple strands of evidence, including archaeological survey and excavation, ethnoarchaeology, experimental replication, and archaeometallurgy. Intensive surface survey located concentrations of manufacturing byproducts (i.e. slag) on surface that represented potential production areas. Stratigraphic excavation and subsequent archaeometallurgical analyses of physical remains have been combined with ethnohistorical and ethnoarchaeological data, as well as comparative analogy, to propose a model for prehispanic copper production among the Tarascans. The goal of this analysis is to gain insights into the nature of metal production and its role in the major state apparatus. Although small in scale, this study provides valuable insights into the development of technology and political economy in ancient Mesoamerica and offers a contribution to general anthropological theories of the emergence of social complexity.

Problem definition and research design issues

Prehispanic Mesoamerican societies were characterized by a highly complex organization with relatively simple technologies. Although the region experienced major transformations in culture, especially in its social, political, and economic institutions, technology proved surprisingly persistent. A reason for this may be that the prestige system, as a regulator, played a key role in the functioning and dynamics of ancient Mesoamerican technologies. In western Mexico, scholars believe that the development and control of copper and bronze metallurgy was a critical factor in the consolidation of the ruling dynasty in the Tarascan Empire (Gorenstein and Pollard 1983: 115-116; Pollard 1993: 102, 1997: 741, 750; Weigand 1982: 2). Nevertheless, the data on metallurgical processes and the organization of the industry are extremely fragmentary. Two major problems stem from this central issue: first, the knowledge of the technology itself is incomplete, and second, nothing is known about its associated resource control and contexts of production. Understanding Tarascan metallurgy and its social context, however, requires not only the acquisition of new information, but also the use of an adequate theoretical framework that encompasses both the technical and cultural aspects of this important industry.

Methodology

The present study was formulated with the above concerns in mind. The initial goal of this project was to examine the technology and organization of metallurgy at a Tarascan locale, relating the evidence for the size and organization of the industry to the character

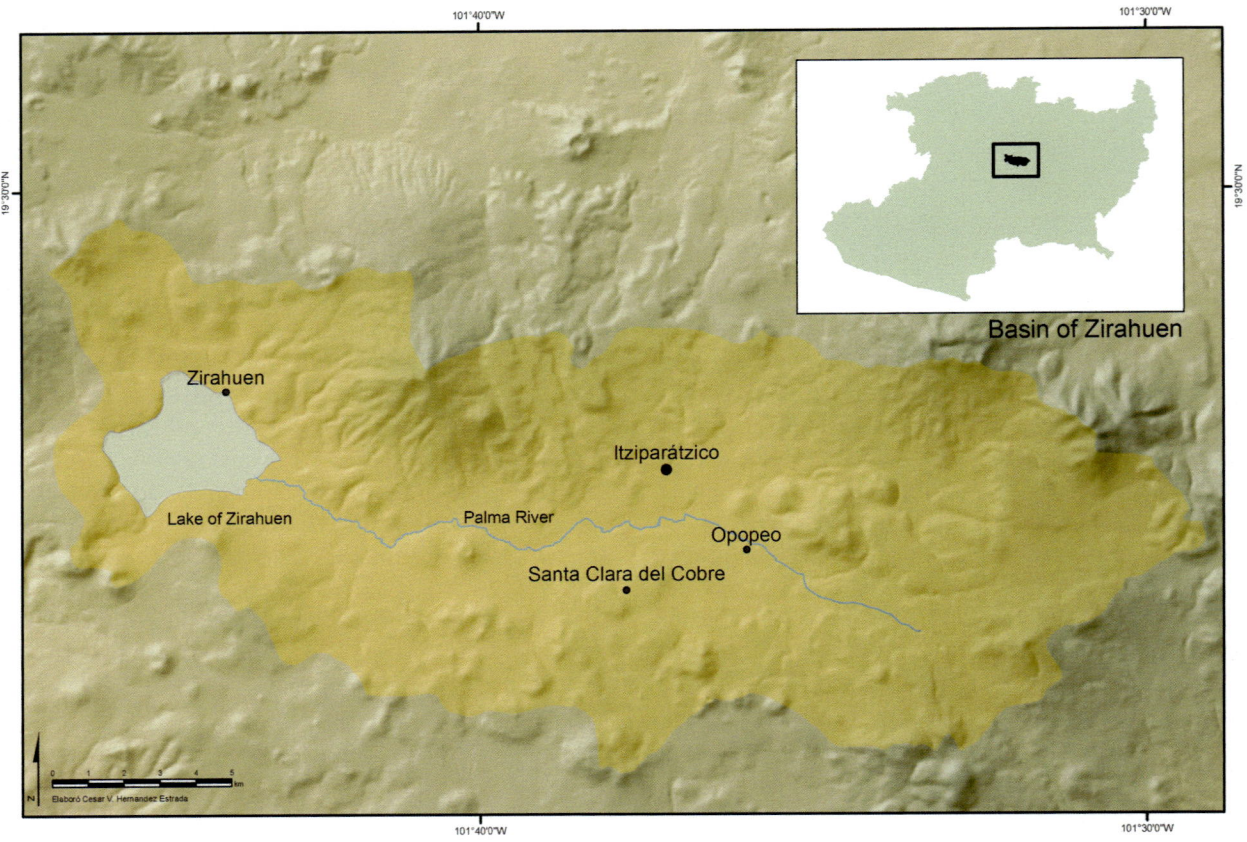

Figure 1.1 Location of Itziparátzico within the Zirahuén Basin and the state of Michoacán

of technological processes, resource requirements, demand, and potential and actual production levels. These goals were addressed in a program of surface survey, test excavation, and laboratory study of the archaeological evidence for early metallurgy at Itziparátzico, Michoacán.

This archaeological zone is located between the communities of Santa Clara del Cobre and Opopeo, in the Zirahuén Basin, in the modern state of Michoacán, Mexico (Figure 1.1). This research program was preceded by preliminary ethnoarchaeology and experimental replication at Santa Clara del Cobre (Maldonado 2001, 2002, 2005) and by the investigation of relevant ethnohistorical sources. An effort was made to systematically collect data on the history of metallurgy in the study area, although the ambitious character of these goals was tempered by limitations in time, personnel, and money.

Data collection and sampling

With the support of FAMSI and the Pennsylvania State University, I initiated systematic archaeological investigations at Itziparátzico during the summer of 2003. Intensive surface survey was used to locate production areas represented by slag concentrations. GIS technology was employed to map an area of approximately 15sq km and record archaeological materials and features. Archaeological test excavations were conducted over a nine-week period in three major sectors of Itziparátzico. Seven test-pits were excavated through deposits of silt and clay in different parts of the site to confirm the existence of *in situ* craft activity and to obtain stratigraphic samples of material remains to date metallurgical production deposits. Regrettably, no identifiable metalworking structures (furnaces, hearths, and pits) were found during the test-pitting. Slag samples recovered from the excavation, therefore, represent the most relevant data for the purposes of this project. While these smelting waste products were recovered in large amounts, only a representative sample of 2.1kg was selected and exported for metallographic analysis.

Significance

Western Mexico has been largely overlooked in ethnohistorical or archaeological investigation. Recent research, however, has helped to define the characteristics of civilization as it appeared in the region. One particular aspect of West Mexican culture that has been studied is the development of metallurgy, in which the recent work of Dorothy Hosler (1985,

1988a, 1988b, 1994) on metalworking technology, and that of Dora Grinberg (1990, 1996, 1997) and others (e.g. Roskamp 2001; Warren 1968, 1989) on ethnohistorical documents, have been of great influence.

Tarascan studies have also played a major part in West Mexican studies. A great deal has been learned about the Tarascan culture in recent years, in good part from the research of Helen Pollard (1983, 1987, 1993, 1997) around the Pátzcuaro Basin, in Michoacán. The purpose of my research is to link multiple lines of evidence, using a comparative approach, to produce as complete a picture as possible of prehispanic copper production in ancient Michoacán. The results of this research represent a contribution to Mesoamerican studies and to the anthropological theory of craft specialization, by leading us to a better understanding of the dynamics of technology and political economy in the Tarascan state.

The approach: metallurgy as technology

Metallurgy has often been a source of speculation for scholars of diverse disciplinary backgrounds. Classical and medieval writers were among the first to systematically consider the nature of metals, their value, the means by which they could be won from the earth, and the cultural implications of metallurgy for human society. This interest continued in the thought of the founders of modern anthropology and archaeology. The use of the terms 'Bronze Age' and 'Iron Age' by Thomsen in the early nineteenth century marked a concern with the cultural significance of metallurgy and technology that was maintained in the evolutionary thought of V. Gordon Childe, Leslie White and others in the twentieth century. The very foundations of archaeological classification betray a preoccupation with the significance of materials, particularly that of metals.

The tendency of this thought, at least among anthropologists and archaeologists throughout the nineteenth and early twentieth centuries, was toward a sort of technological determinism. A concern with the material basis of economic and social life, paralleling the development of the dialectical materialism of Marx and Engels, led some to attribute primary or exclusive causal importance to technology. In the nineteenth century the correlation of technological advance with cultural evolution in the archaeological record supported such a view. In particular, the apparent link between the appearance of complex societies in the Old World and the development of metallurgy reinforced the technological determinist viewpoint. Cultural complexity seemed somehow a result of technological advance.

Metallurgy, among various human technical endeavors, has often been given a prominent role in cultural evolution. The common presupposition has been that metallurgy, as an inherently more complex technology than, say, lithics or pottery manufacture, required a correspondingly complex administrative or organizational apparatus (e.g. Childe 1930, 1942). The complexity of metallurgy presumably derived from its more diverse and specialized requirements for natural resources, equipment, and facilities involved in the smelting procedure.

Implicit in this view of metallurgy were models of both the technological process and the organization of technology drawn from modern industrial metallurgical production. Despite ethnographic and archaeological evidence to the contrary, archaeologists have persisted in interpreting the archaeological evidence of early metallurgy in terms of metallurgical processes and technological organization in current or historically known use.

Ethnographic research during the late nineteenth and early twentieth centuries (especially in Africa: see Cline 1937, for example) demonstrated that metallurgy could exist as a fully developed technology in a wide variety of cultural contexts and environmental conditions. Metal production could occur in both tribal and state-level societies as either an occasional occupation for part-time specialists or as a full-time specialization whose production was mobilized for the purposes of the state. Research has demonstrated that industries using similar technological processes can differ widely in the organization and scale of resource procurement, production, and distribution. Chinese iron production during the eleventh century may provide an example of this dynamic.

In eleventh-century China, the development of the iron industry resulted in two contrasting and coexistent modes of production. The first productive mode was characterized by self-sufficient villages operating on a small scale, and was originally typical of all Chinese iron mining and smelting. The second mode of production was based on a large-scale industry organized in terms of state and market demands which emerged during intensive economic expansion in the eleventh century. These contrasting modes of production developed in different areas of China and the differences appear to have been accentuated by geographical isolation. This variation, however, was economic rather than geographical or geological in nature (Hartwell 1962, 1966).

It was (or should have been) apparent from these data that technological innovation and intensification of production occurred in widely varying economic, social, and political contexts. Yet there has been a continuing tendency to interpret the evidence of early metallurgy in terms of models of modern metallurgical production, attributing to these early industries an economic significance analogous to that of modern industries. The problem is similar in many respects to the interpretation of early irrigation technology (cf. Wittfogel 1955, 1957; Sanders and Price 1968; Downing and Gibson 1974; Hunt and Hunt 1976).

In the New World, the development of technology followed its own path, one both similar to and different from that of the Old World. The knowledge of metallurgy and metalworking evolved and spread over much of the area occupied by high civilizations in the Americas. Copper, gold, silver, and their alloys were the main metals involved. These materials were fashioned primarily as ornaments used in religious ceremonies and for the enhancement of elite cultural status. The manufacture of metal tools and weapons was secondary and occurred relatively late.

Rethinking the materialistic approach to history and anthropology along with a great expansion in archaeological data, has led anthropologists to reject technological determinism and the primacy of technology in cultural evolution. But a viewpoint emphasizing the fundamental importance of technology in cultural change persists in non-anthropological thought and even among some anthropologists. This tendency can be seen in our common usage of terms like the 'Space Age' or 'Computer Age'. Few would deny that technological changes (the development of electronics, computers, communication systems, etc.) are important motivating forces behind much contemporary cultural change.

The rejection of technological determinism by anthropological archaeologists and the recognition that technologies are cultural products have led to a more balanced approach to the relationship of technology and culture, an approach emphasizing their mutual influences. It should now be apparent that, as Childe (1944; McNairn 1980: 74-103) recognized, technological change occurs in a context of existing social and economic arrangements and that the causal arrow points in both directions. Something more fundamental than technology is involved in this relationship: energy transformations and the control of energetic transformations represent the basic link between humans and nature, as Richard N. Adams (1975) has noted.

Much of the confusion concerning the role of technology in culture has arisen from the weakness or absence of relevant anthropological or archaeological data. Archaeology, as an anthropological study with a unique perspective on time and cultural change, has much to offer to the discussion of technology's role in culture. The number of archaeological studies concerned with the cultural context of technology has grown in recent years. Archaeological interest in metallurgy, in particular, is evident in the increasing number of volumes reporting field and laboratory study of metallurgy, its products, and its cultural and environmental settings (e.g. Betancourt 2006; Craddock 1980, 1991; Juleff 1996; Lechtman 1980; Raber 1984; Oddy 1977; Wertime and Wertime 1982).

Probably the first and foremost need in the study of technology and culture is to rethink and make explicit the models applied to the interpretation of archaeological evidence for early metallurgy. Assumptions about both the possible technological processes in use and the possible range of organizational arrangements need to be reconsidered and clarified. Preconceived notions about technological processes may seriously color interpretations of the archaeological data and affect our understanding of resource requirements, the level of technical skill required, and production potential. Presumptions concerning the organizational requirements of metallurgy (or any other technology) can seriously overestimate the scale and necessary administrative arrangements of an early industry.

What is needed is a comprehensive contextual approach that links technology, technological change to the natural environment and the resource base and the historical and archaeological evidence for settlement patterns, trade, craft specialization, and political relations. Such an approach would avoid the myopic focus on technological products and the particularistic concern with origins and techniques in favor of a truly anthropological view of technological development.

The obvious problem with such an approach is the large amounts of time, labor, and money required for a comprehensive program of site survey, excavation, environmental study, and laboratory analyses. Ideally, the survey and testing of technological sites would be a part of an ongoing regional project. In the absence of large-scale projects, we must rely on small-scale initiatives focused on specific technologies, which address questions of resource procurement, production arrangements, administration, organizational scale, distribution, and the interaction of technological and cultural processes. In a small-scale project like the one presented here, the option was to incorporate multiple lines of evidence in a site-specific manner and to apply them in a weight-of-evidence approach focused at the regional level.

Specific issues covered in this volume

Several problems, questions, and assumptions concerning the technological and cultural background of copper metallurgy in the area of study and in the Tarascan region as a whole were defined and addressed during the development of the present study. In outline, they are as follows:

1. Theory of technology. The existing theoretical approaches to explore the dynamic relationship between technology and culture were *unsatisfactory* for a proper treatment of a multiple-perspective-based research. The study of the metallurgical industry at Itziparátzico required a theoretical approach that regards technology as an integral and active component of human systems, that interacts with other parts of such systems in diverse ways. The construction of a multi-approach framework in which the different conceptual parts are dynamically interrelated and interdependent was therefore required. This framework is presented in Chapter 2.
2. Archaeometallurgy. Given the paucity of data and limitations of available analyses, the archaeometallurgical record for Mesoamerica is fragmentary and dispersed. Most of the available information on metallurgical processes is largely based on metallographic analyses of finished products (e.g. Hosler 1994) and thus, often restricted to the final stages of production (i.e. fabrication, surface treatment, and finishing of metalwork). Chapter 3 develops a technological framework that encompasses the complete metallurgical operational chain including ore sources, mining technology, mineral processing and extractive metallurgy, and structures the interactions among these aspects of production.
3. 3. Archaeometalurgical Materials. The use of ores is invariably related to the formation of slags because slags act as collectors for impurities introduced into the smelting process (Bachmann 1982). For this reason, the composition and properties of metallurgical slags are influenced by variables such as the ore itself, the fluxes added, the process conditions (heat distribution, air intake, furnace profile and height retention time of slags within the furnace) and cooling conditions, among other factors. The analysis of slag and other smelting byproducts is therefore of particular importance to obtain technological information on the smelting process. No archaeological data on ore processing, extractive metallurgy, and descriptions of slags from Mesoamerican contexts were available until the year 2003, when the present research began. I was aware, notwithstanding, that similar projects were being carried out at other locations in western Mexico (e.g. Hosler 2002; Roskamp et al. 2003). Results of slag analyses from Itziparátzico are introduced in Chapter 4.
4. The Organization of Copper Metallurgy. The interpretation of the archaeological evidence for early metallurgy may be colored by implicit models of technological organization derived from modern industries or the large and well-known state-directed ancient industries of the Old World. Several explicit alternative models should be considered in evaluating the archaeological record. Two generalized models of organization of metals industries derived from ethnographic information are presented in Chapter 5, where the data from the study are considered as they relate to these models. The nature of local production in the region was related to larger economic and political patterns. The scale and the nature of demand and organization of the industry are all subject to the influence of external economic and political factors. A model of the organization of the Tarascan copper industry is developed to discuss the larger historical context of metallurgy in the region.

Chapter organization and content

A major problem encountered during the development of the present study was a lack of a common theoretical framework that could address metallurgy as both a form of technology and a specialist produced craft. *Chapter 2* provides a theoretical background and a comparative perspective from which to draw useful frameworks for analysis. It also offers an overview of definitions, theoretical models and concepts relevant to this project, in order to build a shared language and logic for subsequent chapters.

Chapter 3 introduces some of the key terms and technical descriptions of processes associated with both Mesoamerican metallurgy and preindustrial metallurgy in general. A brief discussion of the emergence of metallurgy in the New World, highlighting its differences and similarities to early Old World metallurgy is also presented. This is followed by an assessment of the evidence on mining and metallurgy currently available for West Mexico, presented in the framework of a hypothetical operational sequence.

Chapter 4 presents a description of the investigations carried out at Itziparátzico and the methods used to analyze the data recovered. This chapter also offers an introduction to the area of study, including background information on both the environmental and historical patterns affecting Tarascan metallurgy and the archaeological interpretations of evidence for early metallurgy. The broad scope of this information is required for two reasons. First, the paucity of information on the region makes it necessary at several points to infer local conditions from general Tarascan or Mesoamerican trends. Second, as noted above, the nature of local copper production is related to wider economic and political contexts and cannot be understood without reference to these external factors.

Chemical data on the slags from Itziparátzico are also provided in this chapter.

Chapter 5 briefly discusses the theory and methods used by archaeologists in the study of craft production, and two possible models of technological organization are presented to interpret the data on copper production at Itziparátzico. The evidence for organizational and technological change in the region's copper industry is summarized and discussed with regard to such organizational models. Based on these two formulations, a tentative model is proposed for the organization of the copper industry in the Tarascan state of prehispanic western Mexico.

Chapter 6 contains the conclusions deduced from this investigation and suggestions for further research. The potential for alternative use of wind-powered smelting at Itziparátzico is also evaluated in this chapter.

Appendix A contains descriptions of the materials recovered from surface and the features recorded during the 2003 survey at Itztiparátzico.

Appendix B is an inventory of the materials recovered from test excavations at Itztiparátzico.

Appendix C presents the results of the scientific analysis of slag samples from Itziparátzico, as well as a brief explanation of phase diagrams.

Appendix D is an extension of Chapter 2, outlining previous theoretical and field studies in order to clarify the various approaches that have been applied to the problem of technology and culture. The effect of implicit or explicit theoretical approaches, including evolutionary theories underlying much of nineteenth and twentieth century anthropological thought, on field and laboratory studies of early or non-industrialized technologies is considered.

Finally, *Appendix E* summarizes the research methods and results of an investigation of the preindustrial use of wind power for smelting of metals, as a potential model for future research in the Tarascan region.

Chapter 2

Approaches to the study of technology and craft production

The dynamic relationship between technology and culture has been discussed repeatedly, not only within the realm of anthropology, but in other fields as well. Archaeology, which is concerned with the reconstruction of past cultures through the retrieval and analysis of their material remains, consequently has to deal with technology and technological change. Due to the diversity of interests among archaeologists and the varied backgrounds from which modern archaeology derives, a wide range of approaches to the problem, both theoretical and methodological has emerged. This variety of approaches or orientations has had an impact not only on the definition of the problem and of technology itself, but also on the development of relevant methodologies to study it. Craft production, while an important aspect of technology, is typically addressed as a completely disconnected issue, rather than being integrated into existing frameworks.

The purpose of this chapter is fourfold: 1) to summarize the historical and conceptual background of the anthropological study of technology; 2) to approach technology from an evolutionary and ecological perspective, which also considers the role of behavior as a mediator between environmental opportunities and constraints (both social and physical) and technological production; 3) to present a multi-approach theoretical frame that encompasses both the most critical mechanisms involved in technological evolution at the behavioral and archaeological scales and the embeddedness of technology in social, economic, political and cultural aspects of the society; and 4) finally, to discuss craft production in terms of its association with technology in Mesoamerica.

The following summary is intended to provide an understanding of the context in which this research project developed. It is neither comprehensive nor exhaustive. Although no such comprehensive study exists in the literature, the large-scale evolutionary or ecological perspective on technology adopted here is present in numerous references on the topics of cultural materialism, cultural ecology, and cultural evolution; among them: Harris (1968), Carneiro (1974), Adams (1975), Sanders and Webster (1978), and Price (1982). The emphasis on the behavioral nature of technology found in more recent theoretical frameworks (e.g. Bamforth and Bleed 1997; Bleed 1997; Fitzhugh 2001) is also discussed. The connection between technology and craft is made through Leroi-Gourhan's (1964) concept of *chaîne opératoire*. Finally, the small-scale study of craft production is addressed from a political-ecological approach, which has been adopted in several Mesoamerican archaeological studies (e.g. Schortman and Urban 2004).

The definition of technology in archaeology: historical background

While archaeologists must deal directly or indirectly with technology at some level, there is little consensus among scholars over the precise meaning of this term. Most definitions of technology are either too broad or too narrow to be applied in an evolutionary approach, and the boundaries of the subject tend to be vague and confusing. Discussions of technology tend to define it in historical/particularistic or evolutionary/ecological terms, generally with four points of agreement between them: technology is a cultural adaptation to the physical and social environment; it comprises both artifacts and techniques; it is patterned and culturally transmitted; and it is a means of harnessing or using energy, or transmitting and storing information. Nevertheless, these four attributes are often treated in isolation and not as complementary and interactive aspects of a whole. The essential points of these two definitions can be summarized as follows:

1. The historical or particularist definition of technology characterizes it as '[a body] of skills, knowledge, and procedures for making, using, and doing useful things' or 'the practical arts' (Merrill 1968: 576-577); 'all the technical activities of man' or 'a set of coordinated activities about some central core or subject' . . . by which humans seek to modify or control their natural environment' (Spier 1970:1-2); or, very generally, 'those activities directed to the satisfaction of human needs, which produce alterations in the material world' (Childe 1954).

Studies which use the above definition are historical or particularist in the sense that they treat technological development as a sequence of particular events, either deriving from independent acts of creation and adoption (technological innovation and development) or from an internal logic. Historical/particularistic definitions also figure in functionalist treatments of technology and culture in which technology fulfills some static biophysical function in a culture. The emphasis is on the description, classification and comparison of technologies in different cultures and their synchronic or historical analysis, primarily through the definition of types and attributes (Bleed 1997). This approach proved useful for the recognition of cultural historical markers by establishing descriptive comparability and communication among researchers (Bleed 1997; Dunnell 1982).

Descriptive categories have had other valuable applications. Distribution studies have shown that they can be interpreted as reflections of social patterns (e.g. Longacre 1968). Replication studies (e.g. Flenniken 1981) and microwear (or use-wear) analyses (Keeley 1980) allow the interpretation of attribute variations in terms of function, linking them to other kinds of economic and ecological patterns. This research was very stimulating to the field, but tended to confuse technology with tools and to focus on too narrow a range of behavior in attempts to explain technological patterns. Historical/particularist studies did not encourage the conceptual study of technology itself either (Bamforth and Bleed 1997; Bleed 1997).

2. An evolutionary or ecological definition of technology views it as a cultural adaptation to the environment. Technology, in this perspective, is a subsystem of culture subject to the evolutionary processes of selection and differential survival that affects the reproduction of cultural features and cultures themselves (e.g. Steward 1955; Clarke 1968; Price 1982: 715-718). It is both form and content: both 'an information system which connects the species of biological organism Homo sapiens with its environment' (Dupree 1969: 529-530) and the tools, materials, and processes so organized.

According to Raber (1984: 20-21), a reasonable anthropological and evolutionary definition of technology might include two main features. The first is the manipulation of the material world to effect energy transformations. This aspect of technology represents 'a set of knowledge, skills, and materials (apparatus) necessary to alter the order . . . of some set of energy forms or achieve an energy conversion' (Adams 1975: 14). It is in this sense that technology does 'useful' things or is 'directed to the satisfaction of human needs' (Childe 1954). The second feature in this type of evolutionary definition is the assumption that technology is a subsystem of culture subject to evolutionary processes of adaptation and selection. In this view, the evolutionary analysis of technology thus involves an understanding of the physical environment as it impinges on technical choices and their effects.

Inherent in both the historical/particularist and evolutionary ecological views is the concept of technology as human-generated, and thus the result of human intentions and human decisions. Technology, therefore, is intrinsically behavioral. Steinberg (1975), discussing technological styles, describes technology as a subsystem of culture which interacts in a number of definable ways with other subsystems. Technological analyses should therefore include not only the techniques involved in the manufacture of implements, but also its human factors and its products (Steinberg 1975: 54).

Theoretical approaches to technology and culture

As discussed above, existing studies of the relation of technology and culture can be divided largely into two classes: historical/particularist and evolutionary/ecological. To the former belong most of the descriptive studies of technology or technologies by anthropologists, archaeologists and historians. Technology is viewed as an aspect of culture whose component parts or processes have static functions. The methodological emphasis is on description, classification and comparison.

The evolutionary analysis of technology is represented by the work of those nineteenth (and twentieth) century anthropologists and archaeologists who treated technology as a component of an evolving culture. Culture is viewed, in this approach, as evolving through a succession of states or stages and technology is often seen as the primary means through which cultures adapt or develop. The associated methodology is concerned with the description or analysis of technology in terms of its adaptive value, the environmental factors affecting technology, the ability of technology to effect changes of state in the cultural system, or the causal links between technology and cultural change.

The historical/particularist approach

This approach characterizes both the work of non-anthropologists (historians, prehistorians, classical or historical archaeologists) and many anthropologists and anthropological archaeologists. Such an approach is apparent in historical discussions of 'Technology and Culture', in those sections of ethnographies entitled 'Material Culture' or 'Technology', and in the use by archaeologists of technologies and their products as cultural or chronological indicators. Technology, in this view, is assumed to have a fixed role in society and to change as the result of historical events or through an internal logic of cultural development. It therefore has a characteristic form in each society and may be taken as a marker of a particular culture or phase.

The assumptions behind this perspective are that a technology is the product of distinct events of innovation, borrowing or diffusion of a historical sequence of such events in a particular culture, or of some inherent tendency in the human mind. In the historical view, technological change results from a particular combination of circumstances, events, inventions, and materials and from fundamental human tendencies: growing familiarity with materials, inventiveness, and an inherent trend to complexity. In agreement with Bleed (1997: 96), the most valuable aspect of this approach is in clarifying the distinction asserted between technology ('customs of manipulating the physical world') and material culture ('the products resulting from those customs').

Examples of this approach are too numerous to detail. The bulk of the anthropological, archaeological, and historical literature on technology previous to the 1990's falls into this category. The historical/particularist perspective is evident in general studies like Barnett (1953), Singer et al. 1954-1959, Forbes (1955-1972), and Hodges (1970), as well as in more specialized works like Needham (1958), Aitchison (1960), Lucas (1962), Wertime (1973a, 1973b), Healy (1977), and Wertime and Muhly (1980). Studies of technical processes and products generally have, implicitly or explicitly, a historical/particularist theoretical orientation. These include both descriptive studies of technological products or materials and experimental or replicative studies aimed at the reconstruction of technical processes. Again, the literature is vast: examples include Oddy (1977), Olin and Franklin (1982), Wertime and Wertime (1982), and many of the papers in the Journal of Archaeological Science and Archaeometry.

The evolutionary approach

Technology was an important component of the unilineal evolutionism of the nineteenth century (Harris 1968: 142-179; Service 1971: 5-14; Sanders and Webster 1978). Unilineal theories proposed a general trend through history through greater complexity and the progression of human culture through stages of evolution. Technology was the basis for many of the early unilineal evolutionary schemes and technological criteria and labels were selected to mark the major stages of evolution. The evolutionary stages proposed by Thomsen, Worsaae, Lubbock, Montelius, and others (based on their study of the chronological sequence on northern European artifacts) were technological in character: Stone, Bronze, and Iron Ages marked the successive levels of human development (Daniel 1964: 31-49). Technology was used to describe the evolutionary states. Although technology was used to characterize the progression of culture, no attempt was made to examine the links between technology and the dynamics of culture.

The work of Lewis Henry Morgan and later nineteenth century evolutionists (Tylor, Marx, Engels) was significant in that it expanded the concept of technology to include subsistence activities and other cultural adaptations to the environment as well as tool production; it also gave technology a primary causal role in evolutionary change. In these unilineal schemes technology not only characterized the stages of cultural evolution but, through the agency of technological innovation, was the source of evolutionary change. The technological determinism implied in the evolutionism of Morgan was also sometimes apparent in the work of V. Gordon Childe (e.g. 1944, 1950, 1951a, 1951b, 1954). Childe adopted, with modifications, the stages of Morgan's system (Savagery, Barbarism, and Civilization) and elaborated the technological mechanism behind the progression from stage to stage. Childe's views changed over time and he came eventually to modify the Marxist and technological determinist positions he had taken. However, technology retained a central role in all of Childe's thinking on evolution. Childe's position is examined in some detail in Appendix D, since it defined some of the principal questions in the study of technology.

Childe's position requires particular attention because it is characteristic of much of the nineteenth and early twentieth century anthropological thinking on the relationship of technology, culture, and evolution. It should be apparent that Childe simply offers a possible or plausible link between technology and evolution. The necessary causal links and driving force for that evolution are missing. Childe's work was valuable in offering suggestions as to the potential significance of technology in cultural change although he failed to establish a convincing argument on the causal significance of technology. This is, to a great degree, a result of the misconception of cultural evolution. The crude unilineal view of evolution as a sequence of stages allows for no consideration of the role of adaptation, selection, energetic transformations, and innovation in evolution. Childe's work retains an important place in anthropological thinking as a starting point for an ecological analysis of the role of technology. His examination of the role of technology in surplus creation and the development of specialization formed the basis for considering the relationship of technology to energetic transformations in evolution. This theme was taken up in the work of Leslie White and others (see Appendix D).

Neo-Darwinian approaches

While applications of evolutionary ideas have a long history in archaeology and anthropology, the use of neo-Darwinian approaches in archaeology has been emphasized only in recent years (Lyman and O'Brien 1997). Two main research programs for the application of these approaches in archaeology fall under the heading of neo-Darwinian evolution: selectionism and evolutionary ecology (Broughton and O'Connell 1999). The former branch of this evolutionary emphasis can be traced to the work of Robert Dunnell (1978, 1980), and advocates a strict Darwinian-selectionist interpretation of evolution. Evolutionary ecology has, on the other hand, focused distinctively on the interaction between evolutionary forces and ecological variables in the development of specific adaptations (Winterhalder and Smith 1992). One of the major applications of neo-Darwinian ideas to archaeological studies of technology can be found in the use of material culture to establish relationships between populations of artifact-makers. Material culture is seen as central to the human phenotype and, therefore, subject to pressure from either 'selection' or 'adaptation' (Braun 1990, Kuhn 2004; Leonard and Jones 1987, Leonard and Reed 1993, O'Brien and Holland 1990, 1992, 1995).

Three basic tenets underlie many of these evolutionary approaches (Jones et al. 1995; Teltser 1995). First, variation is inherent within the population and this variation is expressed as the phenotype. Archaeologically, the expression of the phenotype represents one of a range of potential morphological forms. Within a particular class of artifacts (e.g., ceramics, lithics), variation in the morphological form is largely continuous, but tends to be classified into more or less discrete categories based on arbitrary divisions of this continuous variability. For example, the division of lithic artifacts into categories of flake, blade and bladelet is based on metric attributes and dimensions that are arbitrarily designated as meaningful and used to partition this continuous variation. Second, these variants are transmitted from individual to individual in a non-genetic fashion. That is, the variants are learned from person to person and transmitted across space and time. Thus, the potential rate of change is much more rapid than in the genetic transmission of traits. Third, among the range of variants within the population, some occur more frequently than others as the result of a sorting-out process analogous to natural selection.

Variants that are present in the population and passed down from individual to individual are changed in their frequency on the basis of the selective advantage they confer. This process of selection assumes that the traits adopted have some sort of selective advantage. If traits are neutral with regard to selection, then any apparent trends favoring a particular phenotype are expected to be the result of stochastic processes.

From an evolutionary point of view, the persistence of a particular phenotype is thought to imply some sort of selective advantage over other phenotypes. From this view, the advantageous phenotype is the end product of the interaction between the organism and the social and physical environment such that as the environment changes, the phenotype may no longer confer an advantage relative to alternative phenotypes.

Archaeologically, the outcomes of the selection process can, in theory, be monitored by examining the persistence of the phenotype (artifact) in the archaeological record. The phenotype should persist only as long as it is selectively advantageous within the constraints of the local environment (Jones et al. 1995; Teltser 1995). By arguing that 'replicative success' of artifacts (which can be documented by long-term trends in the archeological record) implies that technological systems have undergone evolutionary selection, proponents of this approach have focused renewed attention on the study of technology (Bleed 1997; Kuhn 2004). This proposition can certainly be disputed but that dispute is peripheral to the present discussion.

Behavioral and economic factors

A number of recent archaeological discussions aimed at developing theoretical means of explaining variability and change in material systems are identified with the 'behavioral' research program, which builds on the work of Schiffer (e.g. 1976, 1996) and colleagues (Schiffer and Skibo 1987, 1997; Schiffer et al. 2001). Although marked by theoretical differences (see O'Brien and Holland 1995, Schiffer 1996), behavioral and evolutionary perspectives on technology have actively pursued inquiry into basic causes of artifact variability. Both have emphasized the behavioral nature of technology. Both have also advocated development of theoretical means of explaining material variability and change in terms of ideas that have proved compelling in biology and other fields. Sometimes there is an obvious tendency for the two approaches to coalesce (Bleed 1997; Broughton and O'Connell 1999; Schiffer 1996). Evolutionary ecology, for instance, has developed a subset concerned with explaining behavioral variability called behavioral ecology (Krebs and Davies 1993; Smith and Winterhalder 1992). The majority of applications of evolutionary theory in archaeological studies of technology have derived from behavioral ecology, as well as from costly signaling theory (Zahavi and Zahavi 1997), dual inheritance theory (Boyd and Richerson 1985) and other relatively established bodies of thought applied to phenomena that played out over brief time frames and that are independent of changes in human biology. Many of the concepts used, including optimality models, are ultimately derived from economic theory (e.g. the marginal value theorem, Charnov 1976).

In summary, it is clear that anthropologists and archaeologists have found the problem of technology's role in cultural evolution a persistent and significant concern. A potentially fundamental or causal importance for technology has been frequently proposed but often misunderstood. The technological determinist argument has been adequately rebutted, but the exact significance of technology in cultural evolution remains unclear.

Methodological approaches to technology

The methodology adopted in the study of technology follows the theoretical perspectives of the researcher conducting it, however implicit or explicit. There is, as has been repeatedly stressed, no research without some theoretical bias, regardless of how conscious the researcher may be of theoretical underpinnings of research. The methodologies of technological studies in archaeology fall into the categories of historical/particularist and evolutionary/ecological approaches. Almost all archaeological studies prior to the 1990s belong to the first category. A historical or particularist viewpoint, although rarely articulated, lies behind most studies of ancient technical processes, the history or 'evolution' of technologies and technological organization.

The examination of technical processes, through analytical or replicative studies, has produced a massive literature on the materials and techniques of archaeologically or ethnographically known technologies. Many preindustrial/prehistoric technologies have been studied and their products and processes systematically analyzed and described. These include studies of metallurgy (Coghlan 1942, 1951, 1956, 1972; Desch 1936; Tylecote 1962, 1976; Tylecote et al. 1977; Wertime 1964, 1968; Wertime and Muhly 1980) as well as ceramics (Lucas and Harris 1962; Noble 1965 Olin and Franklin 1982; Shepard 1940, 1956; Tite et al. 1982; Wertime and Wertime 1982) and lithics (various contributions in Hirth 2003; Keiller et al. 1941; Renfrew et al. 1966, 1968).

The requirements of a methodology appropriate to an ecological/evolutionary perspective are still a matter of debate. The use of such a methodology is not apparent in the archaeological study of technology. Encouragement for an ecological approach and methodology are to be found (for metallurgy, ceramics, and lithics, respectively) in Lechtman and Soldi (1981), Matson (1965a, 1965b), and a number of works on Mesoamerican and Aegean lithics (e.g. Hay 1978; contributions in Hirth and Andrews 2002; Renfrew 1972; Renfrew and Wagstaff 1982; Spence 1982). The realization of the approach advocated in these works has been slow and is not represented in the bulk of the anthropological literature on technology.

Scholars of ancient technology have also made critical reevaluations of technologically deterministic arguments for modeling change (e.g. Lechtman 1973, 1977, 1979; Van der Leeuw 1993). An ongoing debate exists between Sackett (1982, 1986a, 1986b, 1990) and Binford (1968, 1989) in regard to lithic technology. Sackett has steadfastly maintained that functionality and efficiency are not universal concepts, but rather deeply embedded perceptions that can vary from group to group. Research into Inka metallurgy carried out by Lechtman (1973, 1977, 1979) Shimada and his colleagues (Shimada et al.1982, 1983), and Epstein (1993), has brought similar focus to the role of competing ethnic and political ideologies in the development of Pre-Columbian metal industries. The lingering problem is that in most of these studies technology has remained incompletely defined and the details of the mechanisms of technological change are often omitted.

Models derived from evolutionary theory

More recently, a variety of models derived from or related to evolutionary theory have been applied to archaeological studies of technology in order to explain technological variation and change (e.g. Bamforth and Bleed 1997; Bleed 1997; Fitzhugh 2001; Oswalt 1982). These models attempt to provide a means of examining behavioral variation in terms of its material, social or reproductive advantages under a given set of conditions, irrespective of who was actually behaving and how they conceived what they were doing. This same uniformitarian character renders at least some evolutionary models useful in a wide range of contexts. The use of evolutionary or economic models, however, does not assume that people are always constrained to be, or always succeed in being efficient. In fact, it is widely agreed that complete optimality will almost never be achieved, as efficiency in one arena must always be limited by the demands of competing behaviors in other realms. These models simply predict how people would act if they were behaving optimally with respect to a limited and well-defined set of factors (Kuhn 2004).

The application of any technological model, however, requires a clear definition of technology. In an effort to incorporate human behavior into the study of technology evolutionary change in technology, Bleed (1997; see also Bamforth and Bleed), has pointed out the difference between the material and non-material components of technology. Departing from the long established notion of the distinction between technology (customs of manipulating the physical world) and material culture (the products resulting from those customs) Bleed (1997: 97) defines the material aspects of technology as technological results, which include material culture and environmental modifications. The non-material aspects of technology are viewed as the content of technology, conceived as behavioral variables; that is, 'what people know, how they apply their knowledge, and the standards that guide their technological behavior' (Bleed 1997: 99). The articulation of results with the environment has a feedback on subsequent technological content. The discrete consideration of strategic behavior implies human decision-making capacity and adds mobility to the concept of technology.

Elaborating on Bleed's ideas, Fitzhugh (2001) has defined technology as 'the *deployment* of *tools*, towards some *end*, according to *learned* methods' (Fitzhugh 2001: 128). Implicit in this definition of technology are four major components: material, practical, informational, and purposive. The *material* component relates to the physical machinery (simple or complex) or *tools* employed toward some end. The *practical* component pertains to the activities involved in putting tools in motion (*deployment*) toward a particular end. This is the active aspect of technology, related directly to its performance. The *informational* component contains the knowledge (*learned*) base and strategy sets underlying technological performance. The strategic dimension recognizes that there may be more than one path toward an end and that technological deployment in a variable environment might favor variable strategies. Finally, the *purposive* component that entails deployment must be goal-directed (*methods*) in order to be technological. This implies that any goal, however rational or irrational, could be held by technological practitioners. Based on this interpretation, Fitzhugh attempts to explain evolutionary change and to relate variation in the contexts of human decision-making to long-term technological change.

Fitzhugh's scheme, derived from optimal decision theory in evolutionary ecology, focuses on the causes of technological change by considering risk-sensitivity in the production of technological invention. This evolutionary ecological model of risk and opportunity cost establishes expectations for the contexts in which individuals are likely to stick to conservative technologies and when they are likely to be inventive. In simple terms, this model predicts that in situations where individuals have more to gain than to lose by larger variance strategies, they would be more likely to be innovative. Alternatively, where little is to be gained and much lost with higher variance strategies (when people are risk-averse), technological behavior should be conservative and focused on the most consistent (low variance) and highly productive techniques already practiced. The model applies beyond the sphere of subsistence. For example, the difference between the strategies of subordinates and elites in the production of wealth in 'transegalitarian' societies (Clark and Blake 1994; Hayden 1994, 1995) can be understood as a difference in opportunities for improving well-being (see Boone 1992; Fitzhugh 2001). Symbolic currencies (prestige commodities and money) make it possible for economies of scale to operate at several discrete levels of production above the level of subsistence well-being (Fitzhugh 2001).

Applicability of evolutionary models

While the above theory can potentially be adapted and applied to the analysis of technological change (and technological persistence) in Mesoamerica, the main value of this model for the purposes of this study is in its definition of technology. First, it acknowledges the fact that technologies serve a host of purposes and that specific studies necessitate a narrowing of the operational set of given technological systems. Military technology, for example, would do little to address resource variation directly unless social competition was a derivative of this variation (Fitzhugh 2001: 140). Second, from a production/output standpoint, it takes into account the multiple components of technology, including tools, operational characteristics, and strategic variables (see also Bamforth and Bleed 1997; Torrence 1989). Each of these components imposes constraints on technological production. These constraints can be classified into categories related to tool production (tool constraints), technological deployment (operational constraints), and the social context of invention and deployment (strategic constraints). The relative costs and benefits of prehistoric technologies therefore, can only be assessed by situating archaeologically recovered artifacts analytically in the larger technological system in which they would have been used, which necessarily involves human and environmental factors.

The theoretical perspective presented by Bleed (1997) and reinterpreted by Fitzhugh (2001) adds conceptual clarity to aspects of technological practice, and comes close to the goals of this work in attempting to effectively articulate the linkage between technology and its derived craft systems of production. This theoretical and methodological approach offers the potential for studying fundamental social units of production, which can be multi-scalar in nature (including individuals, households, corporate groups, and even small localized settlements), and that can change over time. Implicit in the distinctive features of these units are the diverse relations of production, distribution, exchange and consumption within which they existed. In the particular case of Tarascan copper production at Itziparátzico, the application of this theoretical perspective will provide the link between technology, craft production and political economy. The actual implementation of this theory, however, is best attained using a processual framework. Bleed (1997) describes technology as intrinsically processual, if only because the world exists in a context of time that is continuous and unidirectional. Technological activity, in Bleed's view, must begin at some point, move through more or less discrete steps, and come to some kind of an end. Technological activities are hardly unique in this regard since all cultural behavior must be similarly processual in some terms (Bleed 1997: 101).

Implementing the Theory: The Study of Technological Processes

The processual operation of technology is especially marked and particularly important in determining the form and development of technological systems and should be a fundamental starting point for the study of technology (Bleed 1997, 2001). A number of complex representations of technological processes have been used to understand how material and non-material aspects of technology change and interact with one another as technological operations proceed (e.g. Bleed 1991, 2001; Bradley 1982; Lemonnier 1992; Sellet 1994; Schiffer 1976). These representations or models are useful because they show that technological operations are made up of activities, when something is done, and choice points, when decisions are made about what to do. They also show that technological processes involve dynamic interactions of the variables that make up technology as a whole. Technological activities are most often easier to recognize because they involve directly observable actions. The decisions that link actions together can also be treated as concrete behaviors, observable both in action and in material results. They offer an objective basis for analysis since they are real things we can observe to study the operation of technology (Bleed 1997, 2001). The processual perspective for the study of technology adopted in the present work is partly derived from the *chaîne opératoire* approach developed by Leroi-Gourhan (1964).

Chaîne opératoire: *the concept and its application*

The concept of *chaîne opératoire* was first outlined for archaeology by Leroi-Gourhan (1964), who brought it into Paleolithic research in France. It has been heavily drawn upon by contemporary prehistoric archaeologists studying the dynamic process of tool manufacture and use. *Chaîne opératoire*, translated as operative chain or operational sequence, has been described as, 'the different stages of tool production from the acquisition of raw material to the final abandonment of the desired and/or used objects. By reconstructing the operational sequence we reveal the choices made by ... humans.' (Bar-Yosef et. al. 1992: 511). Accepting that the individuals in a group have a number of raw materials and techniques available to them; 'identification of the most frequently recurring of these choices enables the archaeologist to characterize the technical traditions of the social group' (Bar-Yosef et. al. 1992: 511). Culture is expressed in these choices that are made throughout the operational sequence. This approach contrasts with the typological approach that concentrates on the end product alone as opposed to the whole process of lithic exploitation (Grace 1996). The main efforts at applying this model have been made in studies of stone tool manufacture in Britain (e.g. Edmonds 1990; Scarre 1999) and in North America (e.g. Bleed 2001; Flenniken 1978).

This largely descriptive technique seeks to document not only the complete sequence of steps, but also the physical gestures entailed in acts of production. The technical sequencing is often depicted by flow charts (Figure 2.1a and 2.1b). The execution of some of the production steps may involve conscious decisions by artisans. Others may involve rote behaviors that are consciously learned as an artisan masters a particular craft, but later become highly routinized and largely unconscious (Bleed 2001). Stone tool analysts have long recognized that the manufacture of chipped stone tools is a sequence of operations in which a block of stone is reduced to form desired tools (e.g. Crabtree 1966, 1968; Collins 1973). The term reduction sequence, for example, is used to refer to the sequence of steps involved in the manufacture of a stone tool. Steps might include roughing out a core, removing blades, and then shaping the blades. The strength of the *chaîne opératoire* approach, however, is that it recognizes that the dynamic enactment of the technical process takes place in interaction with static concepts or sets of rules. In fact, if we look deeper, the concept of the *chaîne opératoire* is far more complex than simply the interaction between knowledge and skill. The central aspect of the *chaîne opératoire* is that the interaction between the technical gesture (the manner of carrying the body) is at the confluence of the mind, body, social world, and material world (see Chazan 2004 for a thorough review of Leroi-Gourhan's theories and concepts).

A number of archaeologists have attempted to address the social context of technological knowledge by merging the precision of the *chaîne opératoire* with the concept of *habitus* proposed by the French sociologist Pierre Bourdieu (1977), also concerned with the dialectic between cognized and non-cognized aspects of cultural behavior (e.g. Lemonnier 1989, 1992; Roux 1989; Roux and Matarasso 1999; Schlanger 1994; van der Leeuw 1994). Other scholars have combined an interest with technological sequences with anthropological approaches to style and 'practice theory' into a perspective that has come to be referred to as 'technological style' (see Lechtman 1977, 1993; contributions in Stark 1998). Both of these methods address 'ways of doing' (i.e. technologies of production) taking into account the social and cultural context in which they are mastered and reproduced. As a research method, the *chaîne opératoire* approach seems consistent with, and complementary to an evolutionary framework which emphasizes both ecological and behavioral aspects of technology, and takes into account the social and political foundations of technological practices. It also offers a way to link specific crafts to the technologies of which they are part. Recent attempts to apply this concept to the analysis of metallurgical processes have been made by Ottaway (1994, 2001), who has examined the production

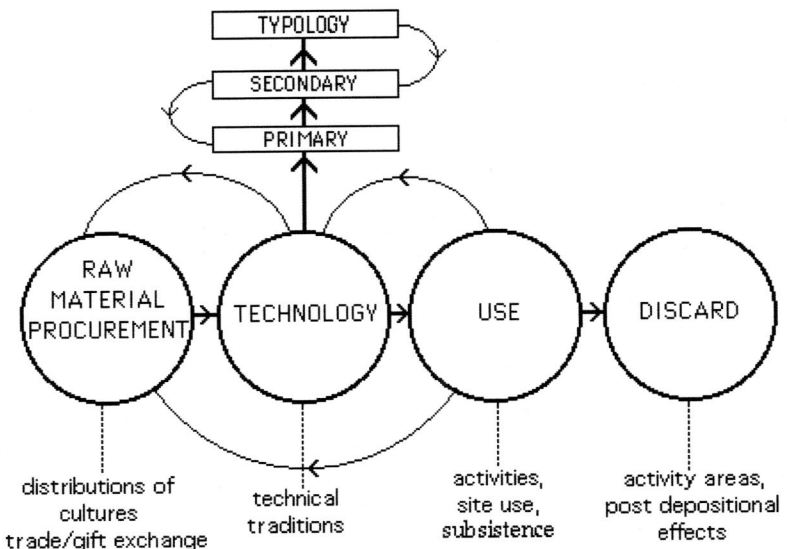

Figure 2.1a The *chaîne opératoire* approach provides a framework capable of linking material and social practices. The four basic links are raw material procurement, technology (separated into primary and secondary reduction and the typology of finished tools), use and discard (after Grace 1996: Fig. 18).

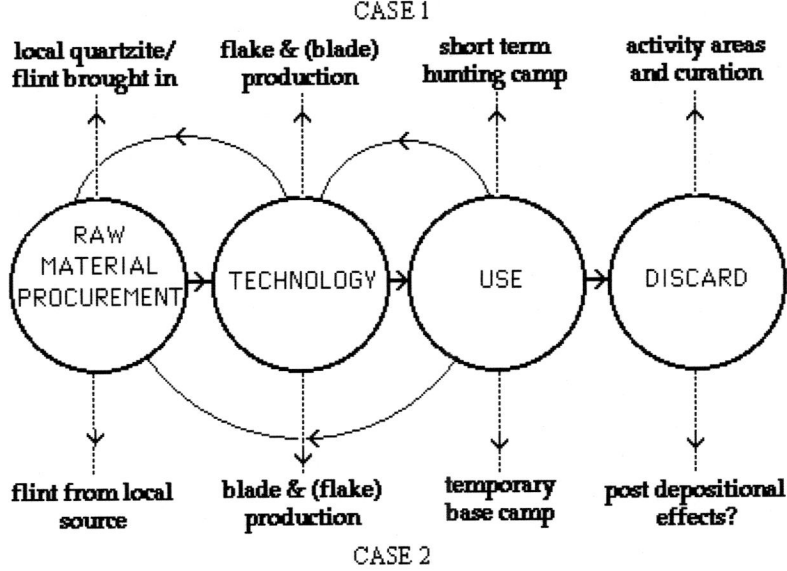

Figure 2.1b The *chaîne opératoire*, however, includes feed-back loops in that the intended use of tools will affect the choice of technology and raw material. If for example the intent was to make projectile points for hunting, this could influence the choice of technology. The figure above indicates and compares the interpretations from two different technological possibilities labeled as Case 1 and Case 2, respectively (after Grace 1996: Fig. 19).

sequences of copper and bronze in early prehistoric Old World metallurgy (see Figure 2.1). Although the European use of *chaînes opératoires* and the American use of 'technical choice' (Schiffer et al. 2001; Schiffer and Skibo 1997; Skibo and Schiffer 2001) are often seen as opposing approaches (see discussion in Loney 2000), they share the basic idea that the way an artifact is made is not a mechanistic process determined by economics, but a fluid one determined by a variety of social factors. Studies using a similar theoretical standpoint, but that do not necessarily use the term *chaîne opératoire*, have also shown that production is social (e.g., Hosler 1996; Lechtman 1977, 1993; Shimada et al. 2000).

Perspectives of the present study: technology, craft production and political economy

Crafts are defined as 'the products of artistic production or creation that require a high degree of tacit knowledge, are highly technical, require specialized equipment and/or facilities to produce and involve manual labor' (*Webster's Ninth New Collegiate Dictionary*). They can be regarded as what in Fitzhugh's definition of technology (Fitzhugh 2001: 128) would represent the *end* products of a particular technology. In this sense, and using Bleed's terminology (Bleed 1997: 99), a craft would embody the material aspects (i.e. technological results) of a certain technology. The study of craft production both as a component, and as the material manifestation of technology, has become central to archaeological inquiry. Technology (as a large-scale process) and craft production (as small-scale or sub-process), however, have rarely been incorporated into a single model. Theoretical frameworks of craft production have been developed independently from theories of technology. The significance of craft production in the formation and dynamics of ancient sociopolitical structures has been one of the most hotly debated topics in archaeology over the last two decades. Issues of power, agency, resistance to domination, and the cultural significance of daily practice that are so pervasive in the archaeological literature, all converge in discussions of specialized manufacture (see Schortman and Urban 2004; Smith 2004).

One of the first efforts to relate a given culture's technological development to its changing political and economic organization can be traced back to the work of Childe, who noted that control of technology was often used as a political tool (Patterson 2005). Specialized craft production played a key role in Childe's views of social evolution (see Appendix D). Though Childe's model is now acknowledged to be overly simplistic, the distinction that he drew between independent commoner artisans (Europe) and client specialists working for elites and state institutions (Mesopotamia) continues to pervade the craft production literature (Wailes 1996). Childe's insights have been taken up and elaborated by scholars seeking to understand the roles that craft-workers played in the construction of ancient societies and the creation of hierarchies in these societies. This reassessment of the relations between specialized manufacture and sociopolitical processes is driven by a growing database and an increasing emphasis on the importance of individuals and factions in the production and reproduction of social and cultural forms (Bourdieu 1977; Brumfiel 1992; contributions in Brumfiel and Fox 1994; Giddens 1984; Patterson 2005; Schortman and Urban 2004; Sinopoli 2003). While the importance of craft manufacture to long-term economic processes is undeniable, the notion of craft production at any scale as a purely economic activity that solely generates surpluses to meet market demands has been challenged. Instead, artisans are increasingly envisioned as having actively participated in the genesis of their social and cultural environments. The ways in which they went about this creative process and the level of freedom they enjoyed in its achievement are matters of intense debate, which has generated voluminous discussion (see Costin 2001 for a recent review of the general craft production literature).

Technology, Craft Production and Political Economy in Mesoamerica

The preindustrial technologies available to Mesoamerican cultures differed greatly from those in the Old World. In terms of the previous discussion, perhaps the most obvious difference is that metallurgy appeared relatively late in Prehispanic Mesoamerica and was never carried out on the scale seen in the Middle East and Europe (Hosler 1988a; 1994). Artifacts of bronze and other metals and alloys served as symbols of rank or contributed to elite self-promotion. Nevertheless, bright-colored feathers, cacao beans and cotton were even more highly valued goods (Arnold 1996; Blanton and Feinman 1984; Blanton et al. 1993; Drennan 1984). Because Mesoamerican societies had no domesticated large mammals and relied on relatively simple technologies, they have been considered low-energy societies (Webster et al. 1993). Mesoamericans, however, built complex systems to intensify agriculture, improving production through irrigation, the construction of terraces on slopes, and raised fields built in swampy areas. The agricultural economy of Mesoamerica provided surplus production that supported the development of a distinctively Mesoamerican social complexity. The Mesoamerican system was linked by continuity in the use of particular materials as items of wealth and standards of value, reflected in economic, social, and political practices. Craft specialists worked fine ceramics, metals, obsidian, jade and other greenstones, and feathers into signs of distinction; scribes, astronomers and calendar specialists developed and recorded indigenous wisdom; and a select body of people who claimed legitimacy in exercising powers of governance consumed these and other forms of 'high culture' (Joyce 2000).

As a historically linked series of socially stratified, economically differentiated, complex societies, the way people were placed in relation to each other was fundamental to the distinctive character of Mesoamerica. Long-distance exchange, one of the practices through which intensive interaction between different peoples within Mesoamerica was fostered, was centrally concerned with obtaining materials used for marking distinctions between commoners and nobles (Hirth 1992). Costume, a major means of marking distinctions between different kinds of people in Mesoamerica, and for signaling the roles of different people, was typically composed of textiles woven of cotton or maguey fiber. Feathers, polished mirrors, and carved greenstone ornaments were all important components of costumes that indicated special status and rank. The Mesoamerican world as a social system was thus formed by its elite prestige system. This world came into existence around 1000 BC, and it developed within the same basic structure for more than two millennia (Blanton et al. 1993). I argue that the prestige system, as a regulator, played a key role in the functioning and dynamics of ancient Mesoamerican technologies.

Approaches to the study of craft production in archaeology

In a seminal work, Brumfiel and Earle (1987) identified three general theoretical approaches to the issue of elite control of craft production: adaptationist, commercial, and political. The adaptationist approach focuses on the adaptation of human groups to their environment (e.g., Redman 1978). Advocates of this theory carried out regional settlement pattern surveys in many areas, and their reconstructions of regional demography and agricultural practices remain fundamental contributions to the economic study of ancient states (Sanders et al. 1979). By focusing on local adaptations, however,

adaptationist scholars minimized the importance of long-distance exchanges and interactions (Feinman and Nicholas 1991). Their functionalist notion that political elites assumed control to manage the economy more efficiently for the benefit of the society as a whole conflicted with more sophisticated social theories. While certain aspects of this approach have been discarded, important economic variables such as the essential role of population pressure in generating agricultural change remain current (Netting 1993; Smith 2004).

Brumfiel and Earle (1987: 1) describe the commercial development model as one in which 'increases in specialization and exchange are seen as an integral part of the spontaneous process of economic growth'. Although cases in which social complexity originated through commercial development may be rare among the earliest states (Trigger 2003), in many more recent instances (e.g., the Swahili and Silk Road economies), commercialization did actually generate social complexity (Smith 2004). In the political model, according to Brumfiel and Earle (1987), local elites assume control of the economy. Unlike the adaptationist approach, elites take a more self-centered stance by strategically controlling aspects of the economy for their own economic and political ends. Since 1987, the political model has developed in two directions (Smith 2004). One approach has emphasized the role of the individual actor, elevating 'agency' and 'practice' to central concerns of archaeological research. Although this view offers some valuable contributions, economics has been pushed aside by political strategizing, prestige, emulation, identity, and gender as major foci of empirical research and theorizing (Pauketat 2001; Smith 2004; Stein 2002). The second research direction to emerge from Brumfiel and Earle's political model is archaeological political economy (Smith 2004).

The Organization of craft production: political ecology vs. political economy

Research into how economic production was organized and surpluses were mobilized in prehispanic Mesoamerican societies has often been approached through one of several political economy perspectives. Anthropological political economy, whose principles are rooted in Marxist thought, has been defined as 'the study of the structural relationships that defined the means of controlling wealth and creating inequality in state-level society' (Hirth 1996: 204). Archaeological studies, however, often focus on aspects of control and inequality, resulting in a more narrow definition of political economy as 'the system that mobilized and allocated the goods and services that funded state activities' (D'Altroy and Earle 1985:189). This is more commonly considered to be the system by which rulers and agents, acting on their own behalf, controlled material (and certain non-material) means of creating and sustaining administrative bureaucracies (Lohse 2004). Variations of this definition underscore the different possible strategies by which elites supervised or directed the production and availability of material wealth, staples and foodstuffs, or ideologically-charged goods in the course of supporting state activities and institutions (Blanton et al. 1996; Brumfiel and Earle 1987; DeMarrais et al. 1996; Hirth 1996; Inomata 2001; Masson and Freidel 2002). Control of intensified food production for example, has often been placed under the auspices of managerial elites (Earle 1997; Ford 1996; Sanders 1977) who, in theory, appropriate surplus production in the form of tribute 'upwards' through bureaucratic levels of administration. This includes mobilizing subsistence goods (Brumfiel and Earle 1987: 4) from sustaining areas, as well as directly supervising and overseeing specialized exotic craft production that occurred within central precincts (see Inomata 2001) and managing long distance exchange networks (e.g. Rathje 1971).

Thus, while Mesoamerican political economies are modeled as consisting of multiple potential strategies for organizing surplus and controlling the allocation of certain goods (Blanton et al. 1996; Earle 1997:7; see Hirth 1996), analytical emphasis is most frequently placed on mechanisms of control and appropriation. This, admittedly, is a simplified summary and a great deal of diversity in terms of how adherents of political economic approaches view a number of important topics can be found in the literature. Key among these is whether the possibility of imposing control over intensified utilitarian production is ceded to non-elites; the importance of local spheres of production in regional politics; and how localized, domestic-scale production is articulated with regional exchange of exotics in sustaining administrative networks (see Schortman and Urban 2004, for an important discussion of these issues). This last factor, how local production combines with extra-regional exchange to sustain ongoing community-level political processes, has been of central concern to the research efforts of the present work.

Rather than structuring the examination of production and exchange patterns of utilitarian goods from fixed natural resources according to a political economic model, my research project on Tarascan copper metallurgy follows a political ecology perspective that lies at the intersection of related fields such as political economy and human ecology (e.g. Ensor et al. 2003; Greenberg and Park 1994; Robbins 2004). Political ecology, too, is partly grounded in a Marxist concept of political economy, but maintains a balanced emphasis on the historical relations of production while also centralizing contested or heterogeneous views of control and social power. This last aspect is perhaps the main element that differentiates the approach from more traditional political economic ones. Political ecology holds two very important implications for how archaeologists model ancient economic production of all kinds, as well as the role of local social non-elites in a given political order (Lohse 2004).

First, political ecologists question some dominant narratives that have emerged from political economic approaches, such as whether elites monopolized or controlled all, or even some sectors of production. Rather, utilizing household studies such as those by Chayanov (1986), Sahlins (1972), Netting (1993), and others, political ecology sees local producers as representing a storehouse of knowledge about how natural resources are best exploited. For example, recent work in the Maya area (Henderson 2003) indicates that some production technologies fell under the control of local householders rather than community or regional elites, which suggests a certain degree of autonomy on the part of common-status craftspersons. Following other researchers (Giddens 1979; Joyce et al. 2001; Paynter and McGuire 1991; Shanks and Tilley 1992), political ecologists reject the normative view of social power operationalized by traditional political economics in favor of a more multidimensional definition that blends 'power over' as the ability to impose order over the actions of others (e.g. institutional representatives directly and forcefully imposing control over resources and having the ability to implement coercive sanctions) with 'power to' as a capacity to transform and express self will (e.g. a more subtle and indirect control over resources, involving inducements and rewards rather than coercion).

Giddens (1979: 6) has termed the balance between 'power to' and 'power over' as the dialectic of control, fundamentally defined as an 'intrinsic relation between agency and power.' That is, it refers to the ability of informed actors to exercise agency, productive power involving local resources for example, in the face of power-over strategies exercised by others (Lohse 2004). In crafting a political ecology framework for the present study, it seems that the manner in which control dialectics were resolved is all too often taken for granted in political economy approaches. This has remained the case even though control dialectics hold important implications for how surplus material goods were made available 'upwards' through social and political hierarchies to support and sustain local, community-level administrative bureaucracies. Strategies by which control dialectics were resolved lie at the very heart of complex, and therefore politicized, social relationships.

Second, fundamental social units of production were likely to have been multi-scalar in nature and to have included individuals, households, corporate groups, and even small localized settlements, which may have changed over time. A good example of the kinds of multi-scaled and interrelated systems of production that this approach envisions is the model of the organization of economic production at Cerén, El Salvador, proposed by Sheets (2000). Ties between producers and local and regional elites were also dynamic, as were the mechanisms by which utilitarian production was integrated into regional economic systems. The political ecology view applied here, therefore, foregrounds the dynamic between centralized control over production and exchange on the one hand and dispersed strategies for managing different localized or dispersed economic resources on the other. Centralizing this process will allow us to better understand the tensions between a variety of different forms of localized producer coalitions and surplus-consuming social elites.

Discussion

A major problem encountered during the development of the present study was a lack of a common theoretical framework that could address metallurgy as both a form of technology and specialist produced craft. Nowhere in the literature is there a systematic understanding or satisfactory approach that encompasses both technology and organization of production, much less one that includes the role of social and political institutions in these interactions. The lack of articulation between technological and craft production frameworks reveals a major problem of scaling issues, the inter-relatedness of patterns and processes at different scales of time and space, to our understanding of technological systems. The principal aim of this theoretical review was to integrate major developments in theory and methods in order to improve understanding of large-scale patterns and processes and their relationship to patterns and processes at smaller scales. The combination of approaches advocated in this study of Mesoamerican metallurgy is ecological and evolutionary (large-scale), and also behavioral and processual (small-scale). Behavior is seen as a mediator between environmental opportunities and constraints (both social and physical) and technological production, thus emphasizing people's diverse strategies for achieving goals or solving problems.

Analyses of technology and technological change should focus on: a) the facets of the environment directly related to technological production; b) the social arrangements facilitating technological processes or necessitated by them; c) the features of technical processes related to resource or energy use and to organization; and d) the changes in cultural relations engendered by changes in technological organization or by the uses of a technological product. The methodology for such analyses should thus include: environmental descriptions and analysis focused on the identification of the pertinent features of a technological adaptation; a regional perspective; and a consideration of the economic and political context of the technology, that evaluates the potential or lack of potential for energy utilization and transformations, changes of scale, economic profit, centralization, specialization, and the consolidation of power.

Technology is an extremely broad and amorphous subject and it is apparent that as an aspect of technology, craft production has the same characteristics. The general goal of this chapter has been to provide a theoretical background and a comparative perspective from which to draw useful concepts and frameworks, i.e. models, ways of organizing ideas and data. Models relevant to this effort derive from behavioral ecology, optimal decision theory in evolutionary ecology, the *chaîne opératoire* approach and political economy/ecology. The previous pages also offer an overview of definitions, concepts and models relevant to our project, in order to build a shared language and logic for subsequent chapters. Basic ideas are presented and the same ideas are then applied to the topic of interest to this study, i.e. copper metallurgy at the Tarascan archaeological zone of Itziparátzico.

Chapter 3

Synopsis of preindustrial metallurgy as applied to Mesoamerica

The study of ancient American societies has often swung between two poles: a search for generalized stages of development in comparison with the Old World, or the perception of New World cultures as unique. Nineteenth century views of cultural evolution did not fit comfortably with societies such as the Maya, which had full written language and sophisticated art and architecture, yet little in the way of metallurgy, or the Inca, who ran a vast empire and had advanced metal skills, but not writing. Pre-Columbian metallurgy was a local development that followed its own path, both similar to, and different from that of the Old World.

Studies of physical and chemical processes are essential to a scientific approach to metallurgy, but because technology is entwined with cultural dynamics, technologies should not be divorced from cultural contexts. The analysis of technology is best approached by a theoretical perspective that regards it as an integral and active component of human systems. Archaeometallurgical research provides a unique opportunity to explore the particular ways in which ancient New and Old World peoples used technology for their own means and ends. This chapter is thus an attempt to provide an introduction to the general concepts and principles of preindustrial mining and metallurgy, presenting it within a comparative framework that emphasizes Mesoamerica and West Mexico.

My research project, although small in scale, represents a pioneer effort to elucidate important aspects of a metallurgical *chaîne opératoire* (Leroi-Gourhan 1963; see Chapter 2) for copper production in Mesoamerica, including ore deposit, mining, smelting, and processing. The study involved the employment of multiple strands of evidence, including archaeological survey and excavation, ethnohistorical investigation, ethnoarchaeology and experimental archaeology, and metallographic analyses of manufacturing waste (i.e. slag). In the following pages, the individual stages from prospecting for metallurgical production to the finished product will be discussed with respect to prehispanic West Mexico. A complete description of the project background, context, purpose, and methods of research is presented in Chapter 4.

Background: the emergence of metallurgy in the New World

The development of metallurgy as a craft occurred gradually over a long period. The time span between the use of metals such as silver, gold and copper in their native state, and the process of roasting and smelting ores of these minerals was extensive. The earliest evidence of smelted copper in the Old World occurred during the sixth millennium BC in the Anatolian-Caucasian region (Craddock 2001; Rothenberg and Merkel 1995). In the Pre-Columbian Americas, it came much later. New World metallurgy emerged in the Andean region of South America between the Initial Period (1800 to 900 BC) and the Early Horizon (900 to 200 BC) (Lechtman 1980). Metallurgical knowledge seems to have spread from the Andean region northwards, as far as Mesoamerica. Archaeological evidence indicates that, while mining practices were rudimentary, metallurgical techniques were complex and advanced (West 1994).

Although New World metallurgy had its own origins, it shared many patterns with the Old World in its subsequent development. In both hemispheres, cold-working was eventually augmented with casting (both mold and lost-wax) and elaborate assembly techniques, including folding, bending, hammering, and the use of solder. Furthermore, the first purposes to which metal technology were applied were not industrial but rather tied to social and ideological concerns. In Central Europe, precious metal jewelry enhanced the status of chiefs and similar social leaders (Renfrew 1979). In the Americas, copper, gold, silver and their alloys were fashioned mainly as ornaments used in religious ceremonies and for the enhancement of elite cultural status; whereas the manufacture of metal tools and weapons was secondary and occurred relatively late (Hosler 1994; Lechtman 1979).

The oldest well-dated Pre-Columbian archaeological site containing metal artifacts is Mina Perdida (Lurin Valley) in coastal Peru, where hammered foils and gilded copper are preserved in contexts dating to 1400 to 1100 BC (Burger and Gordon 1998). The tradition of sheet-metal working (hammering, gilding, annealing and *repoussé*) remained pervasive in the Andes throughout the Early Intermediate Period (200 BC to 600 AD) and the Middle Horizon (600 to 1100 AD) (Lechtman 1979). By 1000 AD, large-scale copper smelting and bronze production is evident at sites such as Batán Grande on the northern Peruvian coast (Shimada et al. 1982). Beginning in the Late Intermediate Period (1100 to 1450 AD), intensive copper-working became widespread on the Bolivian altiplano. The production of materials was based on copper-tin alloys (i.e., bronze), in contrast to the copper-arsenic artifacts found in Peru (Lechtman 2003). By this time, silver and gold were well-established as precious metals among Andean cultures (Quilter 1998).

A second important metalworking center in South America extended from northwestern Ecuador to northern Colombia and eventually into Panama and Costa Rica. The Colombian and Central American smiths concentrated on gold and a gold-copper alloy known as *tumbaga* (Bennett 1946; Hosler 1994). The goldsmiths of the Colombian-Ecuadorian littoral were also the first people to make use of indigenous platinum. The dating of this metallurgy is difficult but probably goes back to the early centuries BC and continues until at least 800 AD (Scott 1992). By AD 90 the craftsmen in the La Tolita-Esmeraldas area were making use of sintering (a type of powder metallurgy process) in the production of their platinum and gold alloys, some of which are plated with platinum over gold by the process of diffusion bonding of small platinum grains to the gold surface (Scott 1992, 2002; Scott and Bray 1980, 1994). This pre-Columbian platinum metallurgy has no parallel in the Old World, where platinum remained unknown until the seventeenth and eighteenth centuries (Scott 1992; Scott and Bray 1980, 1994).

In Mesoamerica, metallurgy and metalworking developed much later than in South America, with West Mexico being the earliest locus of the craft (ca. AD 600). West Mexican metallurgy and metalworking, especially among the Tarascans and their neighbors, was based mainly on copper and its alloys (Hosler 1994: 127; Pollard 1987: 743). The relatively late date of the appearance of metals and the similarity of the techniques employed by the native metalsmiths to those developed in South America has led many scholars to suggest that metal objects and metallurgical techniques were introduced into West Mexico from Peru and Ecuador by traders using watercraft capable of long-distance voyages along the Pacific coast of South and Central America (Edwards 1969; Meighan 1969; Montjoy 1969; Hosler 1994). Hosler (1988a, 1988b, 1988c, 1994) and others (e.g. Arsandaux and Rivet 1921) have extensively discussed the technological relationships between western Mexican and South American metallurgies.

By the time of the Spanish Conquest, three main centers of metallurgical production co-existed in the New World: the Peruvian area, the Colombian-Lower Central American region, and the Tarascan-West Mexican zone (Hosler 1994; West 1994). Native American metal craftsmen from these centers rivaled their European counterparts in the sophistication of their technical skills, but the ideological and social constructs in which they worked were very different. The Spanish were caught up in the early stages of modern capitalism. Conversely, for Native Americans metals were of great value, but were not a commodity, in the Marxian sense, since they were not produced for the purpose of market trade. For some Pre-Columbian peoples, gold for instance was considered the 'feces of the gods' (e.g. Sahagún 1969-1982 book 11: 233), valuable to humans but, ultimately, the waste products of greater truth and beauty. Thus, general statements about the importance of certain metals for a society and about societal developments as a direct result of its need to create institutions and strategies to permit the use of metals cannot be made. Each society must be studied and evaluated on an individual basis.

Mining, metallurgy and the evidence for West Mexico

According to Hosler (1988a, 1988b, 1988c, 1994), metallurgy appeared suddenly in the western region of Mexico (Figure 3.1) between AD 600 and 700. As was the case in Peru, metallurgy and metalworking in West Mexico, especially among the Tarascans and their neighbors, was based mainly on copper and its alloys. Although some utilitarian implements such as needles and fishhooks were made, most metal objects were considered to be sacred, to be used for adornment in religious ceremonies and to enhance the social and political status of the elites (Hosler 1988a, 1994; Pollard 1987, 1993). West Mexican metallurgy thus represents a valuable benchmark for understanding the cultural framework within which it developed.

The production of metal artifacts requires a specific body of knowledge and skills which imply an efficient utilization of what Forbes (1950: 24) has called 'the four essential elements of metallurgy': 1) ores; 2) fuel and fire-making; 3) the production of blast air by draught; and 4) the necessary tools, furnaces and crucibles. Because the transformation of metalliferous ore into finished metal objects involves many individual stages, numerous choices have to be taken along the entire sequence of production. Each stage in the process may influence the final product. In Mesoamerica, archaeological data for either mining or extractive metallurgy are sparse and unclear, and this productive sequence therefore remains fragmentary and incomplete.

Aside from the present work, only two other recent studies report evidence of metal production at West Mexican locations, Jicalán El Viejo in central-western Michoacán (Roskamp et al. 2003), and El Manchón, in the Balsas region of Guerrero (Hosler 2002). The data from both studies are still being analyzed and interpretation is in progress. Samples from Jicalán el Viejo, however, have been made available for study and preliminary results are summarized in Chapters 4 and 5 (see also Appendix C). Few artifacts associated with processing have been recorded. It is thus imperative to carry out systematic research of mining and ore-

Chapter 3: Synopsis of preindustrial metallurgy as applied to Mesoamerica

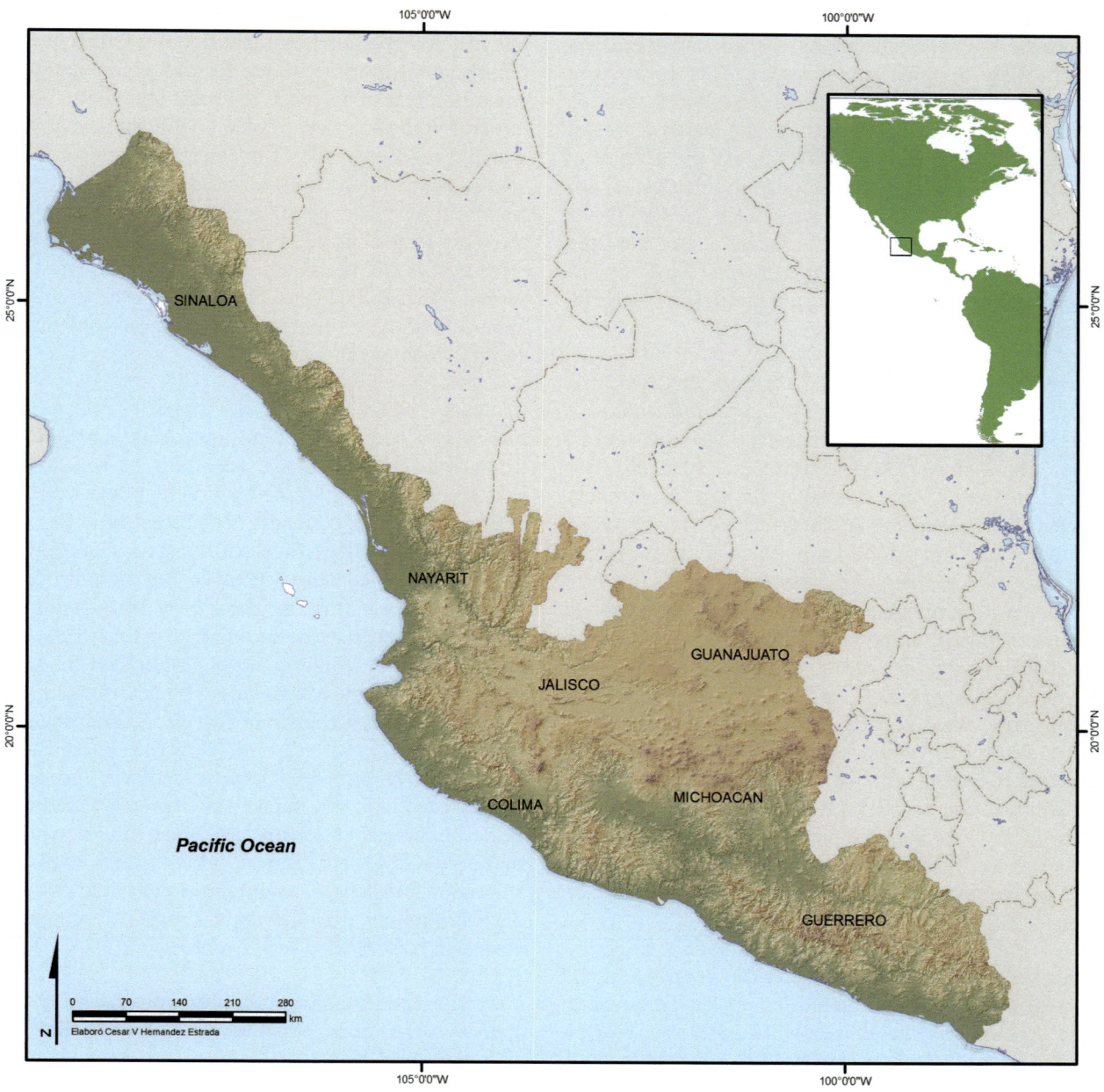

Figure 3.1 West Mexico in the context of Mesoamerica.

processing sites, smelting sites, and metalworking sites to investigate prehispanic mining and extraction in Mesoamerica. Meanwhile, however, the use of multiple lines of evidence including geological, ethnohistorical, and experimental information may provide a basis for minimizing the data gaps in the *chaîne opératoire* of western Mexican metallurgy.

Native metals and ore minerals in West Mexico

A mineral is a naturally occurring solid with a definite chemical composition and crystalline structure that is formed through inorganic processes. All of the ninety-two naturally occurring elements are present in the earth's crust, but many of them at very low average concentrations. Elements such as copper (Cu), lead (Pb), zinc (Zn), and even silver (Ag) and gold (Au) are not rare, but are usually widely dispersed through the rocks. Most minerals present in nature are not composed of a single element, although there are exceptions, such as gold (Au). A mineral is called an ore when one or more of its elements can be profitably removed. It is almost always necessary to process ore minerals in order to obtain the useful element (Forbes 1950; Shackleton 1986).

Apart from comparatively widely scattered deposits of native metals, most metallic minerals found in the earth's crust occur primarily as ores embedded in veins, fracture zones and breccias in mountains

(Shackleton 1986). Exploitable metalliferous mineral deposits, however, are rare geochemical anomalies often with complex mineralogies. This is because most geological/geochemical processes work towards the gradual homogenization of the earth's crustal composition but ore deposits, almost by definition, comprise aberrant concentrations of metals. These abnormal concentrations rarely form as the result of one single process, but normally require multiple events. This means that the majority of ore deposits are geochemically and, therefore, mineralogically variable (Ixer 1999).

Most of the West Mexican territory lies within a metalliferous zone known as the precious and base metal province of Mexico (Hosler 1994; Ostroumov and Corona-Chávez 2000; Ostroumov et al. 2000). The variety of metal ores available in this zone is relatively abundant. These metals and ores include native copper, copper oxides and sulfides, native arsenic, arsenopyrite (FeAsS), sulfarsenides, and different minerals of silver including the native metal, silver sulfides such as argentite (Ag_2S), and silver sulfosalts (Ostroumov et al. 2000). Copper, tin, lead, silver and gold, as well as a number of alloys (see Hosler 1994) were produced in prehispanic West Mexico. Tin deposits are rare, and occur mostly as cassiterite (SnO_2). Apart from some scattered examples in Michoacán, Jalisco, and the state of Mexico, almost all cassiterite deposits are found outside the West Mexican metal zone, in the Zacatecas tin province. Nevertheless, copper-tin bronze became the predominant alloy in West Mexican metallurgy (Hosler 1994).

From the standpoint of Mesoamerican peoples, copper was the most important metal, and played a prominent role in the early metallurgy of Mesoamerica (Barret 1987; Hosler 1994). Copper is a metal more easily smelted than iron and has qualities of malleability, reasonable tensile strength, and corrosion resistance that made it commonly used in pre-industrial times (Forbes 1950). West Mexican copper occurs mainly in the form of ores that had to be extracted from underground deposits. These can be divided into two groups: the easily reducible oxide and carbonate ores, and the more complex ores of the sulfide type. Copper oxides and carbonates are found mainly as cuprite (Cu_2O), malachite ($Cu_2CO_3(OH)_2$) and azurite ($Cu_3(OH)_2(CO_3)_2$); while copper sulfides include chalcopyrite ($Cu_2Fe_2S_4$), chalcocite (Cu_2S) and bornite (Cu_5FeS_4). Added to these is native copper, which is found in a pure state and was readily available on the surface in some locales (Holsler 1994; Ostroumov and Corona-Chávez 2000; Ostroumov et al. 2000).

The metalliferous zone where West Mexican metallurgy developed consists of two geographically distinct mineralization sectors: the southern deposits of Michoacán, western Guerrero, and the state of Mexico, and the northern deposits of western Jalisco, Colima, and Nayarit, with the former region being the copper-richest of the two. The most abundant copper mineral in the southern region is sulfide chalcopyrite, which is present in all of the deposits in Guerrero and southwestern Michoacán. Malachite, which is often a weathering product of chalcopyrite, is also frequent at these deposits. Chalcocite, another sulfide, also appears often at these mines, as well as several carbonates such as malachite and azurite. In the northern region the most common ore minerals are the same as those in the south (Hosler 1994).

Grinberg (1989) has analyzed copper ore from two mines in Michoacán – Churumuco and La Verde – using absorption methods. According to her results, the samples from La Verde contain high levels of iron and copper; other elements are present only in trace concentrations, suggesting that the ore is chalcopyrite or bornite. In contrast, most of the samples from Churumuco lack iron, or show only trace amounts, which suggests that the ore may be either malachite or chalcocite (Grinberg 1989; Hosler 1994). More recently, Hosler and Macfarlane (1996) carried out lead isotope studies of copper ores from several deposits in West Mexico, Oaxaca, and Veracruz, as well as a representative sample of copper artifacts recovered from excavations at a number of sites in West Mexico and other areas in Mesoamerica. Their results indicate that only a few West Mexican mining zones, Inguarán and Bastan in Michoacán, and the regions of Ayutla and Autlán in Jalisco, provided copper metal for a number of sites in West Mexico and beyond.

After the Spanish Conquest, the principal source of copper in New Spain was the Province of Michoacán, with the main center of copper mining at Real de Inguarán. Sixteenth century documents report prehispanic mining activities at this location (Barrett 1987). Inguarán is part of the mining region of La Huacana, in southeastern Michoacán and represents the most important mining center for the purposes of this work. Further details concerning La Huacana are provided in Chapter 4.

Prospecting and mining

By the Formative Period in Mesoamerica, prehispanic societies of almost all levels of complexity had begun to exploit alluvial, placer, sedimentary and hard-rock deposits for non-metal minerals (Weigand 1982). However, apart from references to several mines in Spanish Colonial sources (Grinberg 1995; Hosler 1994; Pollard 1987; Roskamp 2001, 2004; Warren 1968), little is known about copper mining in Mesoamerica. Archaeological investigation of native mining techniques has dealt more with the extraction of nonmetallic substances such as cinnabar and turquoise, than with ores such as copper (West 1994). A major factor in this situation is that modern exploitation of the ore has obliterated most of the evidence of mining and metallurgical activities. Nevertheless, the use of combined data from mineralogical, ethnohistorical and archaeological sources can provide valuable information on mining methods and the ores or minerals mined in West Mexico.

Mining is the extraction of valuable minerals or other geological materials from the earth, usually from an ore body or vein. The first step in mining is prospecting, which involves searching for exploitable mineral deposits and requires some knowledge of the ores, as well as the morphology of the strata in which they occur (Forbes 1950: 49-50; Ottaway 2001: 90). Three general techniques used by preindustrial societies to recover metals following initial exploration have been identified in the archaeological record: surface collection, open-pit mines, and underground mining (Forbes 1950; Ottaway 2001).

The least difficult mining technique in antiquity was *surface collection*, where the ore was available on the surface either in streambeds or exposed on the ground. The erosive power of streams broke up the ore and the heavier metals settled to the bottom in areas of slower flow. These are called placer deposits (Duncan 1999; Forbes 1950; Ottaway 2001). Grinberg (1996: 432) observes that while copper sulfides are a rather unappealing gray in color, when exposed to the surface they weather to form bright colored carbonates such as malachite (green) and azurite (blue). The distinctive colors of these minerals may have served as a guide to prehispanic West Mexican prospectors to track minerals to their source ores.

Where the ancient miners recognized metal ores on the surface, they could follow them into the ground by strip-mining the surface (the debris of undermined mountains), or digging short tunnels. This technique is called *open-pit mining* and is also known as open-cut (e.g. Forbes 1950) or opencast mining (e.g. Duncan 1999; Ottaway 2001). The term open-pit is used to differentiate this form of mining from extractive methods that require tunneling into the earth. Open-pit mines are used when deposits of useful minerals are found near the surface. Where minerals occur deep below the surface (where the overburden is thick or the mineral occurs as veins in hard rock), underground mining methods are used to extract the valued material. Open-pit mines are typically enlarged until the mineral resource is exhausted (Duncan 1999; Forbes 1950: 50; Miller 2003: 102-103).

The third technique was *underground* or *deep-vein mining*. In this case, tunnels were excavated into the rock to remove the ore. Narrow vertical shafts were driven through the rock, widening out to horizontal galleries where the ore was found. Ancient miners preferred to drive adits (nearly horizontal entrances to a mine) or tunnels into the rocky slopes of a valley over digging shafts, as this made drainage and haulage much easier (see Agricola 1556 in Hoover and Hoover 1950: 101-105). The earliest method of underground mining was pitting, involving exploration of the ore-bodies with shafts at intervals which were abandoned one after the other as the ore was extracted from the bottom of the shaft. This method was commonly used by early flint miners in the Old World (Forbes 1950:50).

In Mesoamerica evidence of underground mining of cinnabar, including sizable adits, shafts and galleries dug with hafted hammerstones and dating to AD 1, has been recorded in the Sierra Gorda of Querétaro (Langenscheidt 1970, 1985, 1988). While the mining styles may have varied, similar findings have been reported for the turquoise mines of Chalchihuites in Zacatecas, corresponding to the Classic Period (AD 200-900) (León-Portilla 1978: 12-14; West 1994:13), and for the obsidian mining at Sierra Las Navajas (also known as the Pachuca obsidian source) attributed to the Aztec III (1450-1525) and Aztec IV (1525-1550 AD) periods (Cruz and Pastrana 1994; Ponomarenko 2004). Obsidian mines, however, had no adits.

During the 1940s Hendrichs (1940: 315-16; 326; 1943-44: I, 194 ff.) located a number of open-pit mines in western Guerrero, consisting essentially of large holes dug into hillsides in order to follow oxidized veins of copper (malachite, azurite, cuprite). Evidence indicates that the tools used to excavate the mines and extract the ores consisted of stone hammers, probably made of dense lithic materials such as diorite and andesite and hafted with wood. Hendrichs also reported the presence of large stone mortars, either portable or fixed on the walls of the mines. Other implements include bone scrapers and digging sticks, ceramic ladles, obsidian blades, and wooden wedges. Remains of *ocote* torches and vegetal fibers impregnated with resin, baskets, ropes, and ceramic pots have also been recorded. Unfortunately, no systematic investigations of these features have been carried out in Guerrero in more recent years.

Ethnohistoric documents from early Colonial times in Mexico represent a significant source of data with regard to mining and metallurgy. One of the best-known and most widely studied ethnohistorical sources on Mesoamerica is the *Florentine Codex*, a manuscript containing a hand-written version of the encyclopedic account of Aztec society assembled by Fray Bernardino de Sahagún (1969-1982). Beginning in the 1540s, Sahagún first asked questions of groups of Nahuatl-speaking elders (presumably all male) from the heart of the former Aztec empire. The earliest version of his work, the *Codex Matritense*, consists of short entries, accompanied by drawings, which has become something of an analytic dictionary of terms in the Nahuatl language. *The Codex Matritense* includes references on prehispanic mining such as the designation for mine *in tepeio, in oztoio*, 'lo del monte, lo de la cueva' (that which is in the mountain, that which is in the cave). The term for digging up a mine was *tlallan oztotataca*, which literally means 'excavar cuevas en la tierra' (to dig caves in the earth). The word for copper or metal is *tepoztli*, and *tepoztli iohui* is the term for copper vein (León-Portilla 1978: 11).

Valuable information on mining and metallurgy has also been found in the *Relaciones Geográficas* (Acuña 1985), a group of maps and manuscripts created during the Early Colonial Period in response to a questionnaire developed for Philip II of Spain to survey the diverse regions of New Spain (Mexico), Central America, South America, and the Spanish West Indies during the years 1578-1585. The questions sought information including political jurisdiction, terrain, language affiliation, native traditions, plant names, and mineral sources. The *Relación de Michoacán* (Martínez S. 1903) reports silver and copper mines in areas such as Sinagua and Guayameo, Tacámbaro, and Turicato (near La Huacana), as well as copper mines at La Huacana, all within the territory of the former Tarascan Empire. These documents also indicate that both silver and copper were tribute items in Michoacán. The *Códices de Cutzio y Huetamo* also make reference to towns and villages where metal production took place (Roskamp 2003).

Probably the most famous ethnohistorical source regarding Tarascan mining and metallurgy is the *Lienzo de Jicalán* or *Lienzo de Jucutacato*, a pictographic document dating from the second half of the sixteenth century (see Grinberg 1995; Roskamp 2001, 2004). This document was produced and used as proof of the rights that the indigenous authorities of Jicalán claimed over several mineral deposits, copper sources and soil-based colorants in the *Tierra Caliente* (hotlands) of Michoacán. According to this narrative, the ancestors who founded this town in remote times were Náhuatl-speaking Toltecs who migrated from the east (the coast of Veracruz) towards Western Mexico, where they established the *Cacicazgo* (chieftainship) of *Xiuhquilan* (Jicalán), and began their main economic activities: copper-working and the elaboration of painted gourds (Roskamp 2001, 2004).

Another important sixteenth century manuscript dealing with copper mines in Michoacán is the *Legajo 1204*, published by J. Benedict Warren in 1968. This document narrates the details of a visit to Michoacán by Bishop Vasco de Quiroga during the second half of the year 1533. The purpose of this visit was to investigate the existence and location of copper deposits in the region. The document confirms that several mines were exploited prior to the Spanish arrival, including Inguarán, Churumuco, Huetamo and Bastán, among others (see Chapters 4 and 5). The manuscript mentions other less well known mines, Tancítaro and Coyuca, which apparently were also exploited for copper (Warren 1989; Hosler 1994).

The *Diccionario Grande de la Lengua de Michoacán*, an apparently contemporary document recently discovered by Warren (personal communication 2005) also contains important clues about mining and metalworking in prehispanic Michoacán, including the Tarascan designations for mine (*ynchatzepaqua, haramuta*), copper miner (*tiyamu charapeti haracuquaro*), silver miner (*tayacata haracuquaro*), and gold miner (*tiripeti haracuquaro*), among other relevant vocabulary terms.

Based on her interpretations of indigenous accounts in the *Legajo 1204* (1533), Dora Grinberg has conducted explorations north of the El Infiernillo dam in the state of Michoacán, and confirmed the existence of prehispanic copper mines on the Mayapito hill, near the town of Churumuco. These mines, according to Grinberg (1990, 1996, 2004), are open-pit operations which seem to have been excavated using wooden or antler tools, rather than iron implements. The *Legajo* states that the indigenous people of Churumuco collected green stones from the mines and extracted copper from them. This suggests that the mineral exploited was malachite, a copper carbonate. The presence of this mineral on the surface supports this idea (Grinberg 1990, 1996, 2004).

Accounts in the *Legajo 1204* (1533) declare that the metalworkers from the Churumuco region were actually farmers who cultivated the fields at the foot of the hill where the copper veins were mined, and produced copper for the *Cazonci* (the Tarascan ruler) whenever he requested it. This suggests that mining and metallurgy represented part-time activities undertaken during the dry season (see discussion on ore deposits and mining in the Tarascan territory in Chapter 4). This may also explain the relative scarcity of copper artifacts in Michoacán, given that copper is the most abundant metal in the state (Grinberg 1996: 433).

During their explorations of the Mayapito hill in search of prehispanic copper mines, Grinberg and colleagues located two colonial mines known as Las Guadalupes, which shared the same mineral veins mined by the indigenous people, but on the opposite side of the hill. Unlike the open-pit prehispanic mines, the Spanish mines are vertical-shaft underground mines. By driving vertical shafts into the soil and tunneling the horizontal levels, drifts and galleries into the hill strata, the

CHAPTER 3: SYNOPSIS OF PREINDUSTRIAL METALLURGY AS APPLIED TO MESOAMERICA

Figure 3.2 Medieval mining techniques showing: three vertical shafts of which the first, A, does not reach the tunnel; the second, B, reaches the tunnel; to the third, C, where the tunnel has not yet been driven. D-Tunnel (From Agricola 1556, in Hoover and Hoover 1950: Book V: 103).

Grinberg (1996) has pointed out that archaeological excavations of collapsed mines like this one would provide important data on prehispanic mining in Mesoamerica. In northern Chile occasional incidental finds of the desiccated body of a native miner killed by a collapse of a tunnel roof have provided some evidence of ancient mining techniques. One of these bodies found in a collapsed tunnel was surrounded by various tools, such as a hafted hammerstone, wooden pry sticks and a slate-bladed shovel with wooden handle (Bird 1979). Although our knowledge of Pre-Colombian metallurgy is still fragmentary, far more investigations have been undertaken on native metalworking than on mining techniques in the Americas (West 1994).

Extractive metallurgy

The earliest metals collected by West Mexicans were probably native metals (copper, gold and silver), which occur as such in nature and could be processed into objects by techniques of hammering, tempering, cutting and grinding. The exploitation of mineral ores, however, led to a total change of the methods of metallurgy, including the reduction of ores and the creation of alloys (Grinberg 1996; Hosler 1994). Although the types of metal artifacts and the manufacturing methods and materials (pure metals and alloys) employed in their fabrication have been relatively well established (e.g. Grinberg 1990, 1996, 2004; Hosler 1988a, 1988b, 1988c, 1994; Pendergast 1962; Rubín de la Borbolla 1944), the technological processes used to extract metal from ore remain poorly documented for most parts of the New World. Some general principles of metallurgy, with emphasis on extraction and fabrication of copper metal, are provided below, in an effort to partially fill this void.

Beneficiation

Mineral processing or beneficiation is the practice of separating ores such as oxides, carbonates, or sulfides from waste (e.g. silica and silicates), and represents the first preliminary step in extractive metallurgy. The beneficiation of the extracted ores from mining in preindustrial metallurgy usually involved crushing them into particles that could be separated into mineral and gangue (waste rock), the former suitable for further processing or direct use (Henderson 2000: 220). The process of beneficiation would done either by simply picking out the richest mineral manually (dry beneficiation) or by swirling the crushed material in water and letting the denser ore separate out (wet beneficiation) (Craddock 1991: 61; Ottaway 2001: 92).

In particular cases the beneficiation process is followed by roasting, that is, heating the ore to decompose unstable components, which are bonded to the metal (Cottrell 1995: 99). Evidence of beneficiation activities

Spaniards were able to reach the sulfide ores, instead of extracting carbonates as the natives did (Grinberg 1996: 434). The colonial mines in Churumuco resemble those illustrated in Agricola's book, *De Re Metallica* written in 1556 (Figure 3.2) (see Agricola 1556 in Hoover and Hoover 1950: Book V, pp. 103).

Grinberg (1990, 1996, 2004) believes that after their arrival, the Spaniards took notice of the mining activities carried out by local communities and then, employing European techniques, excavated a shaft or gallery to lower the levels of the same vein. This correspondence between prehispanic and colonial mines was also observed by Grinberg at La Verde, a mine in a neighboring region, which consisted of a small open-pit operation with evidence of colonial modifications. Other prehispanic mines were found in the same mining district, including a very deep open-pit mine located between two hills near the coast (Cerro Camacho and Cerro del Huaco), which is partially collapsed (Grinberg 1996).

was found at the Mayapito mines, where Grinberg and her team recorded the presence of large stone mortars (locally known as *ticuiches*) and pestles presumably used to grind minerals. A feature identified as a grinding table was also located near the mines (Grinberg 1996: 433). These findings may indicate the use of a dry beneficiation method at Mayapito.

Pyrometallurgy

Extracting copper usually involves a chemical reaction by which the metal is separated from the other elements in the mineral. Following separation and concentration, metallic ores had to be subjected to a process of pyrometallurgy (combustion by fire) in which their metallic elements were extracted from chemical compound form and refined of impurities (Grinberg 1996: 436; Ottaway 2001: 93). A pyrometallurgical treatment of ores would reduce their minerals either to raw metals or to intermediate compounds for further refining (Habashi 1986:3, 2005: 165; Tylecote 1986: 12). The basic procedure used in this process is known as smelting, and consists of heating an ore with a reducing agent (often charcoal) and purifying substances to separate the pure molten metal from the waste products (Craddock 1991: 63).

In pyrotecnological processes such as pottery-making and metallurgy, the atmosphere in which combustion occurs is critical. The two opposite ends of the spectrum of combustion are reduction and oxidation. A reducing atmosphere occurs when the oxygen is insufficient for complete combustion to take place. Conversely, in an oxidizing environment the draft in a furnace is strong enough to provide more air than necessary for the fuel to burn. The excess oxygen will combine with any other suitable substance present, in this case, the metal being smelted.

When ores are metal oxides, it is the oxygen that must be removed, and this process requires reduction. The most common preindustrial method of extracting metals from their ores is chemical reduction by carbon or carbon monoxide (CO). Charcoal is composed largely of pure carbon, the other elements having been burned off in the charring process. When burned, it produces quantities of carbon monoxide gas and creates what is called an oxygen starved environment. A reducing environment is ideal for smelting because the carbon or carbon monoxide derived from it removes oxygen from the ore to release the metal (Craddock 1991; Horne 1982; Tylecote 1986).

The smelting process of oxidized ores is relatively simple, consisting of three basic steps: grinding the ore, mixing it with crushed charcoal, and heating it in a crucible or furnace (Grinberg 1996: 436). With carbonates, a temperature of 600-700° C transforms the ore into copper oxide, releasing carbon dioxide (CO_2). If copper oxides are used, a temperature of around 900° C is required and in this case reduction to metallic copper occurs because of the presence of carbon monoxide (CO) produced during the combustion of the charcoal added in a low oxygen environment (Moreno et al. 2003: 627). During the smelting operation, the charcoal burns to form the intensely reducing gas, carbon monoxide (CO). Permeating through the ore at high temperature, the CO reduces it to molten metal. Charcoal thus serves as both the fuel and the reducing agent (Craddock 1991: 63-64; Tylecote 1986: 16-19).

The melting point of pure copper is 1083° C. When oxidized ores are heated to a temperature in the range of 1100° C to 1300° C, the copper is freed from its compounds and merges into droplets. If the residues created by the molten gangue or slag are thick, rather than free flowing, these droplets are trapped inside and can be recovered only by breaking up the cooled slag and picking out the hardened droplets, called prills. Adding a flux such as iron oxide (FeO), sand (silica, SiO_2) or ash to purge the metal of impurities helps melt the slag so that the copper droplets can sink to the bottom of the furnace while the liquid slag floats on top, where it is removed by tapping it off while hot, or breaking it away when it cools (Bachmann 1982: 9-10; Tylecote et al. 1977: 305, 1986: 22). The ore usually contains either quartz (silica) or iron minerals as the gangue, and often both. These two components react together to produce a liquid slag of iron silicates. Therefore, if the ore is rich in silica then iron oxide minerals would be added as the flux; if on the other hand, the ore is rich in iron, crushed quartz would be added as the silicate flux (Craddock 1991: 64).

Early copper workers would discover and exploit the oxidized ores first (since they are found at the top of a mineral deposit, as a result of atmospheric oxidation or weathering of the sulfides) and later, start to work on the sulfidic ores as the easily workable oxides and carbonates gave out (Tylecote 1962). Extracting metal from sulfidic ores, however, involves a much more complicated pyrometallurgical process. Carbon is not a good reducing agent for sulfides, because carbon sulfide (CS) is unstable. It is necessary then to change the sulfides into oxides first (Cottrell 1995).

The reduction of copper sulfides begins with roasting the ores in a highly oxygenated environment to eliminate the sulfur, which gives rise to the formation of copper oxide with an abundant release of sulfur dioxide (SO_2). Roasting needs only an open fire and uses relatively little fuel because the temperature in the first stage must be relatively low, around 600° C, so that the copper oxide that is formed will later react with more copper sulfide to produce metallic copper with the release of sulfur dioxide (Moreno et al. 2003: 628).

The second stage of copper production from sulfides is the smelting of the roasted ore. Roasting in an open container over a wood fire, as explained above, is an oxidizing process. The aim is to remove the sulfides from the ore and replace them with oxides. However, the problem in smelting is one of reduction, which aims to eliminate oxygen, not to add more. In a simple model of copper smelting, restricting the draft would create a reducing atmosphere, but it would also probably lower the furnace temperature below the point where smelting takes place. The solution would be to switch to a reducing fuel, and this is where charcoal comes in once again. For smelting metals, including copper, charcoal appears to be the better fuel. Charcoal burns hotter and cleaner than wood and has the additional side benefit of assisting the flow of the molten metal. Charcoal is usually 85%-98% carbon, the rest consisting mainly of ash, which coincidentally has fluxing properties (Horne 1982).

One type of sulfidic ore, chalcopyrite (copper iron sulfide $CuFe_2$), melts at 880° C and it was discovered that after suitable beneficiation and roasting it was possible to produce copper metal in crucibles without strongly reducing conditions (under charcoal) and without a formal furnace, but at temperatures of c. 1250° C (Henderson 2000; Rostoker et al. 1989). Depending on the raw material and smelting method used, this part of the process will result in the production of: a) copper prills; b) black copper, which is iron-contaminated copper from smelting copper oxide ores using a flux; or c) matte (mixed iron and copper sulfide). All of these products had to be further processed, from simply re-melting to refining, in order to prepare the copper for the next stage in the process. Once the desired refinement has been achieved, copper must again be heated to temperatures of about 1100° C for the molten metal to be cast into a mold. After cooling it may be filed and cold hammered to remove rough edges and to harden it (Forbes 1950; Henderson 2000).

The results of qualitative and quantitative compositional analyses performed by Hosler (1994) on a substantial collection of metal artifacts suggest that the copper used by West Mexican metalsmiths was derived from a variety of sources, including native copper and very pure copper oxides and carbonates, as well as from minerals such as chalcopyrite, arsenopyrite and impure oxides and carbonates. Apparently, the same types of ores were exploited to manufacture a number of alloys. More recently, Hosler (2002) has reported evidence of copper-smelting activities at El Manchón, in Guerrero, involving the processing of malachite and cuprite. Grinberg (1996) had previously examined slag from various Tarascan sites and found evidence of the smelting of sulfidic ores. Slag samples from Itziparátzico have also been analyzed for microstructure and compositional properties and ore being processed has also been shown to be a sulfidic ore, chalcopyrite (Maldonado et al. 2005).

One of the major challenges in the development of metallurgy in the New World was attaining sufficiently high temperatures in the reduction process to smelt metals and their ores. In the ancient Old World, metalsmiths often achieved high temperatures in small furnaces with the aid of hand-operated bellows that supplied a blast of air to increase the amount of oxygen into a mixture of ore and burning charcoal. This instrument, however, was unknown in the Americas prior to the arrival of the Europeans. Andean smiths in Pre-Columbian Peru developed the blow tube made of a hollowed stem of cane through which a worker would blow to produce an air blast directed toward the burning coals inside a clay furnace. A ceramic nozzle, or *tuyère*, was placed on the furnace end of the tube to protect it from the coals (Figure 3.3). The temperature attained in this process was high enough to smelt ore. Scores of broken nozzles have been found by Shimada

Figure 3.3 Experimental smelting at Cerro Huaringa, Peru, based on archaeological data from Batán Grande (Photo courtesy of Izumi Shimada, Carbondale, Illinois).

Figure 3.4 Mesoamerican crucible and blow-pipes (from the *Lienzo de Jicalán*, in Roskamp 2004: Figure 6).

Figure 3.5 Mesoamerican crucible and blow-pipe. Jerónimo de Alcalá, Relación de Michoacán, Moisés Franco Mendoza (estudios y coord. de la edición), Zamora, Gobierno del Estado de Michoacán Colegio de Michoacán, 2000, lám. XIX [primera ed. ca. 1539-1541].

and colleagues (e.g. Shimada et al. 1983; Shimada and Merkel 1991) excavating the important metallurgical site of Batán Grande in northern Peru.

Etnohistorical evidence reveals that a similar form of blow-pipe was used by Mesoamerican smiths. The *Lienzo de Jicalán*, for example, illustrates metalworkers from Xiuhquilan crouching in front of a brazier, melting metal by blowing through pipes (Figure 3.4) (Grinberg 1996; Roskamp 2001, 2004). Similar representations are shown in other documents, such as the *Relación de Michoacán* (Martínez S. 1903) (Figure 3.5) and the Codex Mendoza (Berdan and Anawalt 1992) (Figure 3.6). The actual smelting operations, however, are more likely to have been carried out in pits dug into the ground. References to such facilities are found in the *Relación de Michoacán* (Martínez S. 1903: Lámina XXIX; Warren 1968). When Spaniards took over the local copper industry of Michoacán in 1533, natives were employed as both miners and smelters, and for decades continued to use the metallurgical techniques that they knew. Eventually, however, blowpipes were replaced by animal skin bellows (Barrett 1987: 15, 26, 64). The results of my research suggest that an alternative pyrometallurgical method (i.e. wind power) might have been used by Mesoamerican peoples as a supplementary way of smelting complex ores. West Mexican smelting techniques are further discussed in subsequent chapters.

Alloy technology

Between ca. AD 650-1200 or 1300, West Mexican metalsmiths appear to have worked almost exclusively

Figure 3.6 Mesoamerican crucible and blow-pipe, from the Codex Mendoza.

with native copper and easily smelted oxidized copper ores. But from 1300 to the Spanish Conquest in 1521 they combined copper with other elements to produce a variety of alloys (Hosler 1994). An alloy is a substance composed of two or more metals, or sometimes a metal and a non-metal, which have been intimately mixed, often by fusion. The term is usually reserved for those cases where there is an intentional addition to a metal for the purpose of improving certain properties, such as hardness, malleability, toughness or color, not found in the pure metals. An alloying element, which may be effective in as little as a tenth of one percent, may be viewed as either dissolved in, or in chemical combination with, the major metal. Hence, the metals in an alloy are usually not separable. An alloy with two components is called a binary alloy; one with three is a ternary alloy; one with four is a quaternary alloy (Cottrell 1995: 189; Tottle 1984: 39; Tylecote 1986: 29).

It is debatable whether the earliest alloys were intentional or accidental (Budd and Ottaway 1995; Ottaway 2001). Almost all copper ores contain some small proportion of arsenic, tin, zinc, antimony, or nickel, which can mix at the atomic level with the copper during smelting. It is possible, therefore, that early alloying was not deliberate (Tylecote 1986). Nevertheless, ancient metalsmiths soon came to associate specific mixtures of ores in the furnace with particular results. In time, a skilled smith would have some control over the end product, manufacturing not a random copper alloy, but a specific alloy in which the properties of copper were enhanced. (Ottaway 2001; Tylecote 1986). The intentional alloying of copper with another mineral enabled the craftsman to choose the most desirable property, be it a more silvery color, a metal which could fill the mold more readily, a metal that could be work-hardened to a higher degree without cracking, or one that gave a specific sound or color that was culturally desired (Hosler 1995)

In Mesoamerica, copper was not only the predominant metal, but also the most extensively used base material. Metalworkers in Michoacán, Jalisco, Colima, northwest Guerrero, and the southern parts of the state of Mexico produced an assortment of copper-based alloys, including binary alloys such as copper-silver, copper-gold, copper-arsenic, and copper tin, and ternaries like copper-silver gold, copper-silver-arsenic, copper-arsenic-antimony, and copper-arsenic-tin. Copper-silver alloys were derived mainly by smelting copper ores and silver ores separately, and then melting the two metals together. This combination is unmistakably intentional because there are no ores that contain both copper and silver in concentrations high enough to yield these compositions (Grinberg 1996; Hosler 1988c, 1994, 1995). Artifact analyses suggest that the copper ore mineral for these alloys was chalcopyrite (Hosler 1988c, 1994, 1995).

Copper-arsenic alloys can be obtained by smelting copper sulfarsenide ore minerals such as enargite or tennantite after initial roasting to eliminate the sulfur. These ore minerals contain copper and arsenic (and in the case of tennantite, antimony at low levels), and the metal extracted from them was an immediate alloy of copper arsenic. The alloy can also be produced by smelting mixed sulfide ore minerals such as chalcopyrite, and those containing arsenic, such as arsenopyrite, or the weathered products of these two co-occurring minerals, which may be malachite with copper arsenate or chenevixite (Lechtman 1985; Hosler 1994). These minerals, however, have not been reported in West Mexican mineralogy. Hosler (1994) has suggested that the most likely method used by prehispanic metalworkers in this region was the smelting of mixed-ore minerals of chalcopyrite and arsenopyrite.

Copper-tin alloys can be produced by the direct smelting of copper ore minerals that contain tin, such as stannite, but there is no geological evidence for stannite in Mexico. The copper-tin alloy, therefore, could have been produced in one of two ways: either by first smelting cassiterite to obtain metallic tin, then adding the tin to molten copper, or by co-smelting cassiterite with copper ore minerals. Tin bronze objects were manufactured only in two areas in the Americas, the southern Andean highlands and West Mexico. In the latter, copper-tin bronzes date to after AD 1200. Apparently, West Mexican metalsmiths made copper-tin alloys through deliberate alloying of copper smelted from chalcopyrite with tin smelted from cassiterite (Hosler 1994). The *Relación de Michoacán* (1540-41) describes and illustrates goods made of copper alloys of gold or silver and copper-tin and copper-arsenic bronzes (Martínez S. 1903: 65, 203, 211).

Fabrication of artifacts

With rare exceptions (i.e. mercury), metallic materials have certain characteristic physical properties: they are lustrous, have a high density, are usually hard (when alloyed) and tend to have a high melting point. Metals are also sonorous, which means that they conduct sound well. However, perhaps the most significant characteristics of metals are their ductility and malleability. Ductility is the ability of a material to deform easily upon the application of a tensile force (e.g. if a material can be drawn out into thin wires it is considered to be ductile); while malleability is the capacity of a metal to exhibit plastic response when subjected to compressive force (e.g. working, hammering or shaping under pressure without breaking). These are the properties which make metals capable of sustaining large plastic deformations without fracture and the ability to be shaped and formed by hammering, casting, forging, forming, bending and coiling (Cottrell 1995: 428-442).

In Mesoamerica, metallurgy first appeared in West Mexico between AD 600-800. The earliest metalworking in Mesoamerica includes objects crafted by cold-work with annealing and lost-wax casting (Hosler 1988b, 1988c, 1994). In this context, the terms cold-working and hammering refer to the plastic deformation of metal while cold. Annealing involves heating the metal after cold-work has reduced its plasticity. Casting implies shaping metal in the liquid state (Hosler 1994).

West Mexican decoration techniques included gilding, embossing, soldering and false filigreeing (Bray 1977; Chadwick 1971; Hosler 1994; Pendergast 1962; Pollard 1987; Rubín de la Borbolla 1944; Warren 1968).

The initial evidence for metalworking in Mesoamerica appears mainly along the western coastal plain of western Mexico. Thus far, the earliest recorded artifact is a piece of sheet metal from Tomatlán (Jalisco), dated to AD 600, or even before (Mountjoy and Torres M. 1985). Also in Jalisco, two lost-wax cast bells dating to between AD 650-750 were excavated at the site of Cerro de Huistle (Hers 1990). Other sites include Amapa, in Nayarit (Meighan 1976) and possibly a number of locations along the Balsas River in Michoacán and Guerrero (Cabrera C. 1986; Maldonado C. 1980).

Bells were the most common artifact cast using the lost-wax technique. Most other objects including needles, tweezers, rings, awls and axes, were cold-hammered. Metalworkers used both methods to fashion ritual objects, but utilitarian items were usually formed by hammering. The primary metal used for the manufacture of these early artifacts was copper, either native, or smelted from copper oxides such as malachite or azurite. With time, however, artisans began to employ a very wide range of metals and alloys for both object classes (Grinberg 1996; Hosler 1994; Pendergast 1962).

Cold and hot working

Copper, like most metals, exhibits a physical characteristic that produces dramatic increases in hardness due to cold work. Cold work involves changing the shape of a metal object by bending, shaping, rolling or hammering. As the metal is shaped, internal stresses develop which act to harden the piece. Heat also plays an important role in the work hardening of a material. When materials that exhibit work hardening tendencies are subjected to increased temperature, they may act as a catalyst to produce higher hardness levels in the workpiece (Cottrell 1995: 405-406).

During the early stages of development of metallurgy in Mesoamerica, West Mexican metalsmiths fashioned copper objects by cold-working them from a cast blank. Although most worked artifacts were ritual and status items, tools were also produced. When metalworkers subsequently began to experiment with copper alloys, the enhanced physical and mechanical properties of these new materials allowed them to elaborate, refine and in some cases redesign the same object types that they had been crafting in copper (Hosler 1994, 1995).

The main classes of Mesoamerican artifacts produced by working metal to shape include open rings, tweezers, axes, needles, awls and sheet metal ornaments. These were fashioned from a cast ingot (or blank) usually by cold-working the material, with intermittent rounds of annealing. However, if the material processed was an alloy with a concentration of tin or arsenic high enough to cause brittleness, hot working or forging was employed. The strength of these alloys and their capability to be hardened by heat treatment allowed the development of new artifact designs. In addition to binary copper-tin and copper-arsenic alloys, metalworkers used copper-silver mixes and ternary alloys of copper-arsenic-tin. The significant advantage offered by blends or alloys over single-component systems is that the properties of the material can be fine-tuned to its application. Changing the elements and their concentrations present in the alloy leads to stronger, more flexible materials. Metalsmiths were able to shape finer ritual/status objects such as open rings and tweezers, and thinner, yet stronger utilitarian implements (Hosler 1994). A few of the most representative prehispanic metal artifacts are described in detail below.

Rings. Small open rings worn as earrings or hair ornaments were produced in large numbers in West Mexico. Mountjoy and Torres M. (1985) report the presence of open rings in household and funerary contexts at Tomatlán. Similar findings appear in burials at El Infiernillo sites and also at Amapa. The burial assemblages at Tomatlán and El Infiernillo consist mainly of rings, which are often found near the cranium of the deceased. These rings vary in diameter (ranging from 1.2-4.0cm) and may have either rectangular or round cross-sections (see Hosler 1994: Fig. 3.7 for example). However, metallographic analyses show that the specimens with round cross-sections were shaped from an originally square cast rod. That rod was cold hammered, annealed though several sequences, and then folded along its longitudinal axis, leaving a central fissure. Before AD 1200, nearly all rings were made of cold-worked copper, but later on they were forged in high-tin bronze (>10% tin). Metallographic studies of the bronze rings with round cross-sections show that while these were made in the same way as their earlier copper counterparts, the use of bronze alloys allowed metalsmiths to produce wider and considerably thinner bands (Hosler 1994).

Rings made from high-tin bronze have been found in burial contexts at Tzintzuntzan (Grinberg 1989: 39; Pollard 1987: 744), Urichu (Hosler and Macfarlane 1996: 1821; Pollard 1995: 42; Pollard and Cahue 1999: 265), Huandacareo (Macias G. 1990:129), and Milpillas, as well as Bernard, Culiacán, and Tuxcacuesco (Hosler 1994: 127). The rings recovered at Bernard are made from a high-tin bronze (Brush 1962). The specimens found at Tzintzuntzan, however, are forged in copper-arsenic bronze and copper-silver alloys (Grinberg 1989). The rings from Urichu burials include high-

Figure 3.7 Replica of a Tarascan spiral tweezer.

tin bronze alloys and mixtures of copper and silver (Hosler 1994; Hosler and Macfarlane 1996). The alloying elements in all these cases, tin, arsenic and silver, appear in concentrations high enough to alter the original color of the copper. Hosler (1994) has pointed out that manufacture of thinner object designs requires tin only in low concentrations, and suggests that that metalworkers may have systematically added tin in high concentrations to obtain a range of golden colors. Silvery-colored rings were made by using high-silver copper-silver alloys.

Tweezers. The earliest West Mexican tweezers (also called beam tweezers) consist of two symmetrical blades joined by a hinge that was fashioned from a continuous piece of metal; they range from 4.5 to 7.8 cm in length (see Hosler 1994: Fig. 5.6). Specimens recovered from burials at early sites such as Amapa and Tomatlán are usually made from copper and shaped by cold work, although some were annealed after fabrication. Metallographic analyses carried out by Hosler (1994) suggest that tweezers were crafted by hammering out the entire form from a piece of copper which was reduced by at least 50% in thickness during the process. The resulting flat product was then placed over a model of wood or other material in the shape of the hinge, and bent double along the mid plane (see Hosler 1994: Fig. 5.10). The hinge was then cold-worked to its final shape, the two blade tips aligned, and any excess metal along the edges cut and abraded away (Hosler 1994:65). Ethnohistorical sources (e.g. Craine and Reindorp 1970) suggest that tweezers were used to remove facial hair, but were also worn by priests and other religious leaders and functionaries.

Simulation studies performed by Hosler (1994) on copper tweezers revealed that cold-worked copper is not an optimal material for fully operational depilatory tools, because it cannot tolerate high levels of stress. However, the use of alloys introduced during the later period allowed greater flexibility in design and considerably higher yield strength. The tweezers that metalworkers were producing at the time of the Spanish Conquest show high quality and technical skill in their manufacture. A new design was developed consisting of a shell with a three-dimensional curvature below the hinge, which gives the blades a shallow domelike appearance (see Hosler 1994: Fig. 5.6). Shell tweezers range in length from 3-7cm (although specimens as long as 12cm have been recorded) and were usually very thin (the blades can be as little as 0.01cm thick). The design of these tweezers required the stiffness and strength of copper-tin and copper-arsenic alloys. Metallographic studies show that shell tweezers contain as much as

Figure 3.8 Tarascan priest wearing a metal tweezer.
From Jerónimo de Alcalá, Relación de Michoacán, Moisés Franco Mendoza (estudios y coord. de la edición), Zamora, Gobierno del Estado de Michoacán-El Colegio de Michoacán, 2000, lám. II [primera ed. ca. 1539-1541].

6% tin. These items appear to have been hot-worked to shape, while specimens with smaller concentrations of tin were cold-worked or annealed (Hosler 1994).

A particular variety of shell tweezers known as spiral tweezers, which are distinguished by four symmetrical spirals emerging from each side of the blades (Figure 3.7), appears only in the Tarascan region of West Mexico. These tweezers were not only wholly functional, but also symbols of rank and sacred power (Hosler 1994). The *Relación de Michoacán* (Martínez S. 1903) describes the activities of the Tarascan chief priest and mentions tweezers as items of ritual paraphernalia and priestly office. This document also recounts that tweezers made of gold were offered as gifts to visiting foreign leaders. The Tarascan chief priests were often represented wearing large spiral tweezers around the neck (Figure 3.8). These ethnohistorical records suggest that tweezers were important cultural symbols among the Tarascans. Archaeological data recovered from excavations at El Infiernillo, Urichu, Tzintzuntzan and Huandacareo support this association. Tweezers appear in high-status burials often placed on the chest area of the interred individual, suggesting that they were worn as pendants. West Mexican tweezers are also found in elite burials elsewhere in Mesoamerica (see Hosler 1994; Hosler and Macfarlane 1996).

Sheet Metal Ornaments. With the introduction of alloys in Mesoamerica after AD 1200 yet another new set of objects for social display began to be produced. Extremely thin cold-worked sheets of copper-silver, copper-gold, copper-silver-gold and sometimes silver, gold or silver-gold mixes were used to create ornamental breastplates and shields, headbands, pendants, earrings, disks and bracelets, among other items. Because of their mechanical strength and resistance to cracks, copper-silver alloys are optimal materials for thin sheet metal objects (Hosler 1994). The *Relación de Michoacán* (Martínez S. 1903) and other ethnohistorical sources (e.g. Warren 1985) reveal that sheet metal objects made from gold, silver and their alloys were common in Michoacán prior to the Spanish conquest. According to these accounts, the Tarascan king wore a *rodela* (a silver disk or shield) on his back to lead his warriors into battle, and was buried with a gold disk on his chest. Hosler (1994:154) analyzed a large disk found in a funerary context (see Hosler 1994: Fig. 5.12). The results indicate that this particular specimen contained 95% silver and 5% copper, and that the metal was subjected to intensive cold-work followed by annealing.

Axes. The *Relación de Michoacán* (Martínez S. 1903) and other sixteenth century documents provide significant

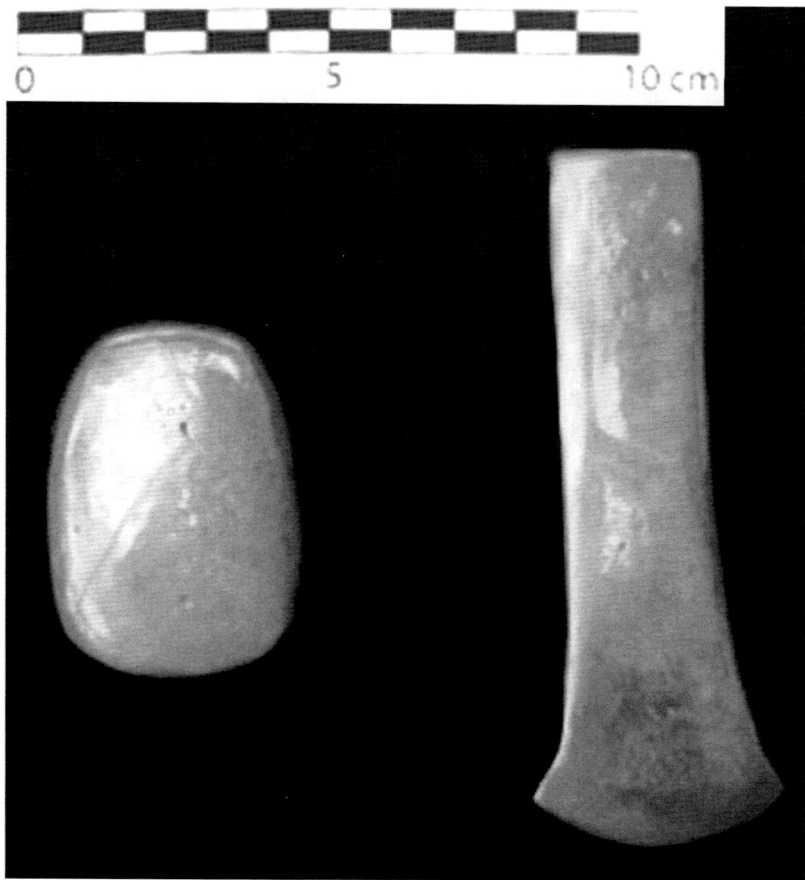

Figure 3.9 Examples of experimental axes cast in copper and tin bronze.

evidence of the use of axes and axe-like cutting implements in Mesoamerica. According to these accounts, axes were used as both cutting tools and markers of social rank and sacred and political power. Axes were made of copper or bronze, their lengths ranging from 8.0-17.4cm (Figure 3.9). Metallographic studies of prehispanic copper axes performed by Hosler (1994: 73) confirm that while some of these artifacts were actually used for cutting, others show no evidence of use and were probably symbolic. Contextual information, while currently unavailable, is critical to fully understand the function of these artifacts. According to Hosler (1994: 75), ritual axes are thinner in proportion to length than utilitarian ones, which made them unusable for most practical applications, since copper is not a particularly strong metal. Microstructural analysis of usable specimens reveals that these items were cast to shape, cold-worked and annealed, and then cold-worked again, a process that hardened the metal.

Copper is not an optimum material for cutting purposes, and even work-hardened copper axes are not strong enough to cut hard woods, although they were perfectly suitable for splitting wood and similar tasks (Hosler 1994: 75). Pure copper (like silver or gold) has a hardness factor of 2.5 to 3 on the Moh's scale, which is about the same as that of limestone. Naturally occurring copper, however, is somewhat harder due to metallic impurities (Tylecote 1962: 42). Old Kingdom Egyptians produced functional copper tools such as chisels, drills and saws. Through tempering, copper implements could be hardened to work freshly quarried limestone from the 4th dynasty onwards, although annealing with fire and hammering also rendered the tools more brittle. Because of the metal's softness, copper tools lost their edge quickly and had to be re-sharpened frequently. Harder stones such as granite, granodiorite, syenite and basalt, could not be cut with copper tools alone; instead they were worked with methods like pounding with dolerite, drilling, and sawing with the aid of an abrasive, like quartzite sand (Arnold 1991).

Axes cast using bronze alloys became common during the Late Postclassic Period in Mesoamerica (AD 1350-1520). Metalworkers employed tin bronze, copper arsenic, and ternary copper-arsenic-tin alloys to enhance the performance of their tools. Axes made from alloys are about three times thinner and the length to thickness ratio increases by about 50%. They are, however, considerably harder than their copper counterparts (Hosler 1994). The Florentine Codex (Sahagún 1969-1982) illustrates one step in the process of manufacturing a prehispanic axe in which molten

Figure 3.10 Casting of prehispanic axes,
from The Florentine Codex (Sahagún 1969-1982: book 11, plate 796).

metal pours out of a crucible and directly into a mold (Figure 3.10). Metallographic analyses show that these axes were initially cast as blanks, and that their final form was shaped by hammering while cold and then annealed. After the final anneal, the axe blades and butt ends were cold-worked once again to harden them. Apparently hardness was controlled by work hardening the finished tool rather than by controlling the concentration of the alloying elements. Standard alloys are difficult to produce and, although metalworkers sometimes did control tin concentrations, they did not actually standardize the alloys (Hosler 1994).

Axe-Monies and other Cold Work Tools. In addition to axes for woodworking and woodcutting, Mesoamerican metalworkers produced a variety of tools by cold-working metals and alloys. These consist of implements for cloth production and metalworking, including needles and awls, and tools for subsistence activities, such as hoes, fishhooks and digging stick points. An entirely new artifact category, axe-monies, was also developed (Bray 1977; Chadwick 1971; Hosler 1994; Pendergast 1962; Rubín de la Borbolla 1944). This artifact type has been reported most frequently in Michoacán, Guerrero and Oaxaca, and was apparently used for tribute. Axe-monies are shaped like an axe, measuring 14-20cm in length, and are very thin (thickness ranges from 0.02-0.06cm) (see Hosler 1994: Fig. 5.27). Their relatively standardized size and portability are consistent with their use as tribute items. Axe-monies are usually made from either arsenical copper or from an alloy of copper and arsenic. However, the concentrations of arsenic vary, which suggests that they were not being controlled in a systematic way. They were cast from an original blank and then shaped by successive cold work and annealing to produce an extremely thin, leaf-like object (Hosler 1994; Hosler et al. 1990).

Lost-Wax Casting

Casting involves pouring molten metal into a mold to obtain a near finished form. Molds for these operations can be made from sand, plaster, metals and a variety of other materials (Cottrell 1995: 183). Cast bells were among the most distinctive metal objects produced in Mesoamerica, frequently found in funerary contexts as jewelry placed around the neck, wrists or ankles of the deceased. In the several documents written around the time of the Spanish Conquest (e.g. the Florentine Codex), bells are represented attached to the garments of elites and deities and to musical instruments. In the early stages of metallurgy, most of these bells were cast in copper. Later on, tin-bronze was also widely used for their manufacture. According to Hosler (1990), bells represent about 60% of the artifacts in West Mexican collections, suggesting that they were produced in larger quantities than any other prehispanic metal object. During the Late Postclassic period, the most important center of production and use of bells was the Tarascan Empire in the Pátzcuaro Basin, Michoacán (Hosler 1994:243; Pollard 1987: 744).

Prehispanic bells exhibit wide variation in terms of size and shape, but they tend to share some general characteristics: they are usually small in size, measuring 1-8cm in height (although much larger specimens have been recovered occasionally). Nevertheless, they all seem to have been cast in one piece, their shape ranging from round to oval to cylindrical; all specimens have a suspension ring on the top and a narrow slit opening at the base, and contain a loose clapper made of metal, ceramic or pebble. Some varieties of bells have extremely elaborate designs, which may have made the casting particularly difficult (Hosler 1990, 1996). During the Late Postclassic Period, bells manufactured through the technique known as wirework were produced in large quantities. These bells appear to have been made from coiled threads of wire forming complex vertical or horizontal patterns, or even recognizable anthropomorphic or zoomorphic figures (Figure 3.11a, b). The original models for these bells, however, were made entirely in wax, by winding pieces of wax thread around a clay core (Grinberg 1989; Grinberg and Franco 1987; Hosler 1996).

Although the manufacturing process of West Mexican copper and copper-bronze bells is unknown, the gold casting lost-wax technique used in the Basin of Mexico at the time of the Conquest is described in detail by Sahagún (1969-1982, book 9: 73) in the Florentine Codex. Some scholars (e.g. Hosler 1996; Long 1964; Torres and Franco 1996) have suggested that the prehispanic technique used to cast copper and bronze bells may have been similar to the gold-casting process described by Sahagún, which involves the following steps:

Figure 3.11a Examples of experimental wirework bells cast in copper and tin bronze.

Figure 3.11b Details of experimental wirework bells cast in copper and tin bronze.

1. The casting of a copper artifact starts with the modeling of a sun-dried mixture of ground charcoal and clay into the form to be shaped. In the case of bells, this initial model should already contain the clapper that will be at the core of the final artifact.
2. Decorative details are added to the created model through the application of small sheets or threads made out of a hardened mixture of beeswax and *copal* (a type of resin) strained and rolled out very thin, and then coated with a thin layer of powdered charcoal.
3. A small wooden stick or vegetable thorn is inserted into the model, leaving both tips visible, which gives stability.
4. A tube or pouring appendix is attached to the wax-covered figure.
5. The piece is covered with a coarse paste made of charcoal and clay and set out to dry. After two days, the model is fired in a crucible or brazier.
6. The space left by the melted wax is filled with molten metal through the pouring appendix and left to cool.
7. Once the metal solidifies, the charcoal and clay mold is destroyed to expose the artifact.
8. The remains of the pouring appendix are removed and any imperfections removed by abrasion or other mechanical means.
9. Finally, the piece is polished and finished.

Long (1964) performed experiments to cast copper bells using the descriptions above. The successful casting of several copper bells supported his hypothesis that the technique used to cast copper objects was similar to what Sahagún described for gold casting. More recently, I performed similar replication experiments to further test and explore this model (Figure 3.12). Nine three-dimensional replicas of bells were produced through a process of trial and error. Three of these pieces were successfully cast in copper-tin bronze and six in pure copper (see Figure 3.11a, b) (Maldonado 2005).

Aspects of this casting technique have also been revealed through micro-structural analyses of West Mexican bell collections. Specimens analyzed by Hosler (1996) present no joints, fissures or porosity, which confirms that the bells were cast in one piece. Remains of the mixture of charcoal and clay from the original mold have been observed in some of the bells found in archaeological contexts. Evidence of the use of wood sticks or thorns is also often present in the form of small holes on the surface of the finished piece (Grinberg 1996; Torres and Franco 1996).

Discussion

Preindustrial metallurgy is often approached as if it were a single technological process. Furthermore, this technological process is frequently treated as if it were devoid of social context. Yet many individual stages are involved and numerous choices have to be taken in the entire sequence of production for the successful transformation from metalliferous ore to the finished metal object. Each stage in the process may influence the final product. In spite of the gaps that remain in our understanding of prehispanic metallurgy in Mesoamerica, the production sequences of copper can be reconstructed and analyzed in a framework capable of linking the different material and social practices involved in metal production. The concept of *chaîne opératoire* provides such a framework by defining a sequence of technical processes, their outcomes, and the choices behind them, but also taking into account the social structure and historical context in which these technical systems were taking place. Figure 3.13 summarizes the operational sequence for preindustrial copper metallurgy as discussed in the present chapter.

Although the precise timing of these technological events and the historical circumstances surrounding them are only fragmentarily known, this hypothetical reconstruction was made possible through the use of

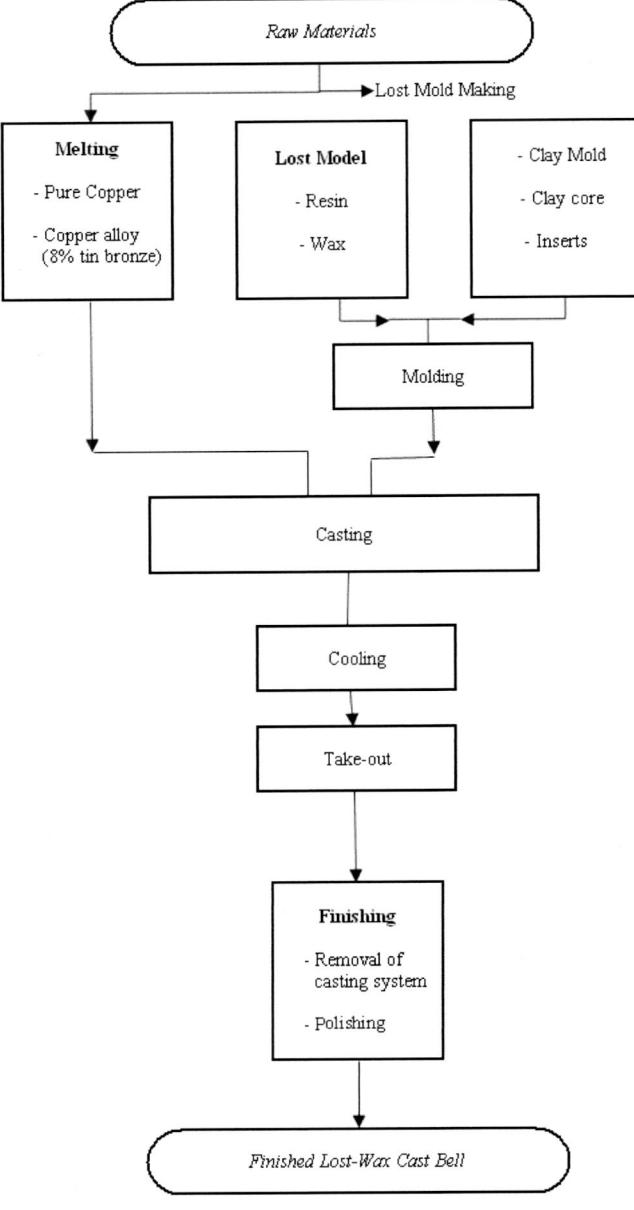

Figure 3.12 Flow chart describing the process of casting experimental copper and bronze bells.

multiple data sources, including archaeological and ethnohistorical evidence, conceptual, scientific and technological analyses, and cross-cultural comparisons. In studying the operational sequence of producing metal artifacts from copper ore, the interplay between technology and organizational forms could be observed from the moment the raw material (copper) had to be procured to achieve the final result: the metal object. As Hosler (1994) has noted, one observable phenomenon in Mesoamerican metallurgy is that during its estimated 900 years of development the technology underwent major changes in materials and methods, but its focus of interest remained unaltered. Hosler (1994) has also

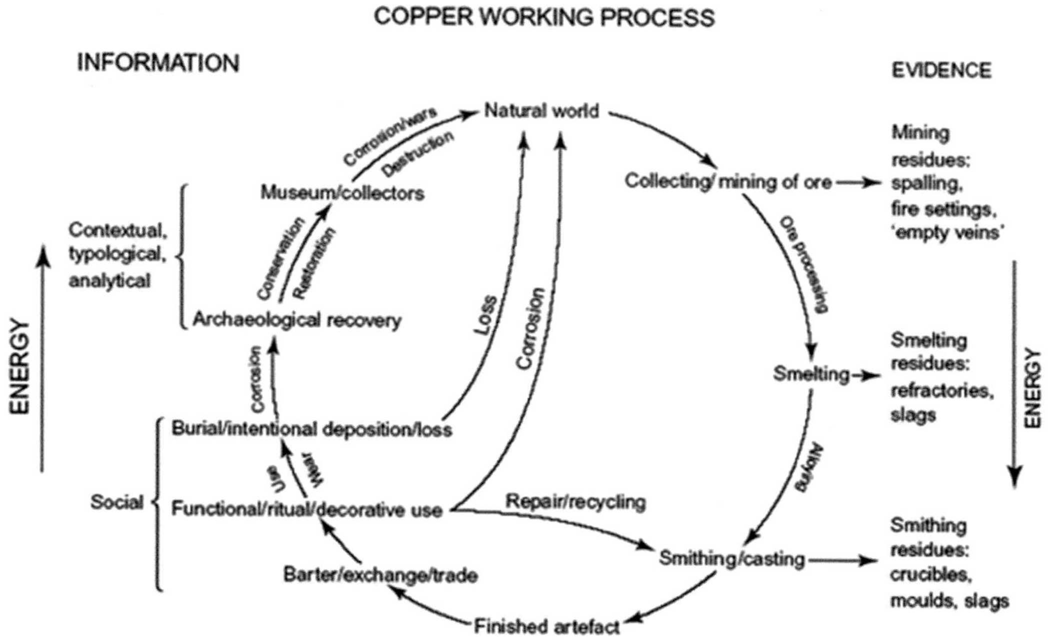

COPPER WORKING PROCESS

INFORMATION — **EVIDENCE**

Figure 3.13 Cycle of copper production and working
(after Ottaway 1998: Fig. 1). Many individual stages are involved and numerous choices have to be taken in the entire sequence of production for the successful transformation from ore to the finished metal object. The concept of chaîne opératoire provides a framework capable of linking the material and social practices involved in this process (Ottaway 1998: 88).

pointed out that, with a few exceptions, the classes of artifacts produced were essentially the same, as was their frequency relative to one another.

The present chapter has been an attempt to move away from technological determinism and look at the concept of metal technology as a social phenomenon. Metallurgy in western Mexico followed its own trajectory of innovations and developments, which could be reflections of social organization or the structure and worldview of local communities, sometimes coupled with environmental factors within that region. At the same time, it is meant to introduce some of the key terms and concepts associated with both Mesoamerican metallurgy and preindustrial metallurgy in general. A brief discussion of the emergence of metallurgy in the New World, highlighting its differences and similarities to early Old World metallurgy has been presented, followed by an assessment of the evidence currently available for West Mexico on mining and metallurgy.

Chapter 4

Tarascan copper smelting in the zone of Itziparátzico: a case study

Over the past few decades, archaeologists have recognized the important role that craft production played in the political economy of prehispanic societies. In western Mexico the development and control of copper and bronze metallurgy may have been instrumental in the consolidation of centralized political power in the Tarascan state (Gorenstein and Pollard 1983: 115-116; Pollard 1993: 102, 1987: 741, 750; Weigand 1982: 2). Mesoamerican metallurgy, however, is poorly documented. Although it has been suggested that metallurgy was under the tight control of the Tarascan elite (Gorenstein and Pollard 1983: 115; Hosler 1994: 155; Pollard 1987: 744, 1993: 119), little is known about resource control and the contexts of production. The present chapter summarizes the results of a pioneer investigation of copper production at a Tarascan locale. The research outlined here examines the technology and general context for copper metallurgy in the Tarascan archaeological zone of Itziparátzico, located in the Zirahuén Basin, between the communities of Santa Clara del Cobre and Opopeo, in the state of Michoacán (see Figure 1.1).

This chapter is organized in three parts. First, a summary of the socio-technological context for copper metallurgy at Itziparátzico is presented, which consists of a brief overview of the Tarascan state followed by discussions of the role metal goods played in the political economy of the state, and their associated patterns of distribution and consumption. Ethnohistorical information on ore deposits and mining, as well as metal extraction, is also provided in this section. Second, a broad description of the physiographic setting highlights the most important environmental aspects of the region surrounding Itziparátzico, the resources, and the history of land use is included. This description is relevant to the present work not only because it places the region in its natural context, but also because the research outlined here represents one of the first systematic archaeological investigation carried out in the Zirahuén Basin as a whole. Finally, I explain the methodology utilized for locating, identifying and mapping the area, the methods and techniques applied in the collection, processing and analysis of the archaeological remains, and the results obtained from this research project.

Socio-technological background: the Tarascan state

In the early sixteenth century much of West Mexico was under the rule of the *Purhépecha* Empire, known to Europeans as the Tarascan Kingdom of *Michuacan* (Figure 4.1). The maximal extent of this political unit was roughly equivalent to the modern Mexican state of Michoacán, but also included parts of Guanajuato, Querétaro, Jalisco, and possibly Guerrero and the state of Mexico (Gorenstein and Pollard 1983:5; Pollard 1993: 5; Warren 1985: 2). Both archaeological and ethnohistorical evidence indicate that during the Late Postclassic Period (AD 1350-1525) this political unit was the primary center for metallurgy and metalworking in Mesoamerica. This technology was largely based on copper and its alloys. Although some utilitarian implements were made, most Tarascan metal objects were considered to be sacred, to be used for adornment in religious ceremonies and to enhance the social and political status of the elites (Hosler 1988a, 1994; Pollard 1987, 1993), thus becoming an integral part of the political economy of the state. For the purposes of this study I define 'political economy' broadly as the relations among political structures and systems and the economic realms of production, consumption, and exchange (Stein 2001: 359; see also Chapter 2).

Metal and the political economy of the Tarascan state

The Tarascan Empire emerged when Tzintzuntzan gained control over five other polities in the Pátzcuaro Basin, as a result of intense competition among local elites over access to basic resources (Gorenstein and Pollard 1983: 118; Pollard 1993: 101-102,177). According to colonial documents, an elite lineage solidified Tzintzuntzan's political control of the Pátzcuaro Basin between AD 1250-1350 (Gorenstein and Pollard 1983: 119; Pollard 1993: 29, 1997: 367). Three additional administrative centers were created, resulting in a highly centralized and hierarchically organized administrative system under the rule of Tzintzuntzan (Figure 4.2). These administrative centers, however, were patronized by resident, non-royal elites (Gorenstein and Pollard 1983: 111; Pollard 1993: 82). Political control was held by a hereditary dynasty and maintained by a vast and complex system of tribute (Pollard 1993: 116). Pollard (1987, 1993, 1997) has suggested that in the process of defining Tarascan elite culture, the central dynasty encouraged the development of metal as a distinctly Tarascan product.

According to Pollard (1987: 741) the rapid expansion of the Tarascan state between AD 1350 and 1450 (see Figure 4.3 for the chronology established for the Pátzcuaro

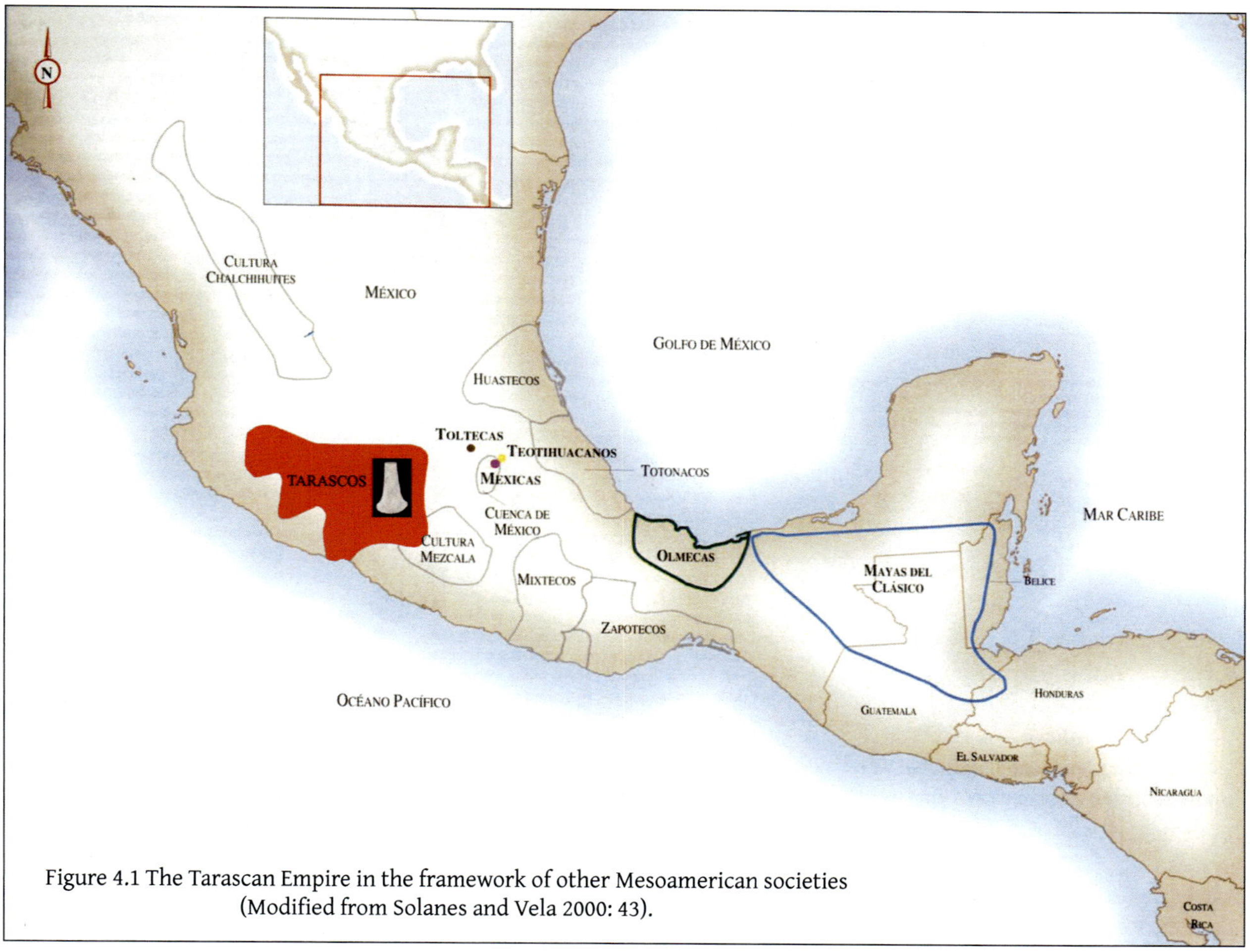

Chapter 4: Tarascan copper smelting in the zone of Itziparátzico: a case study

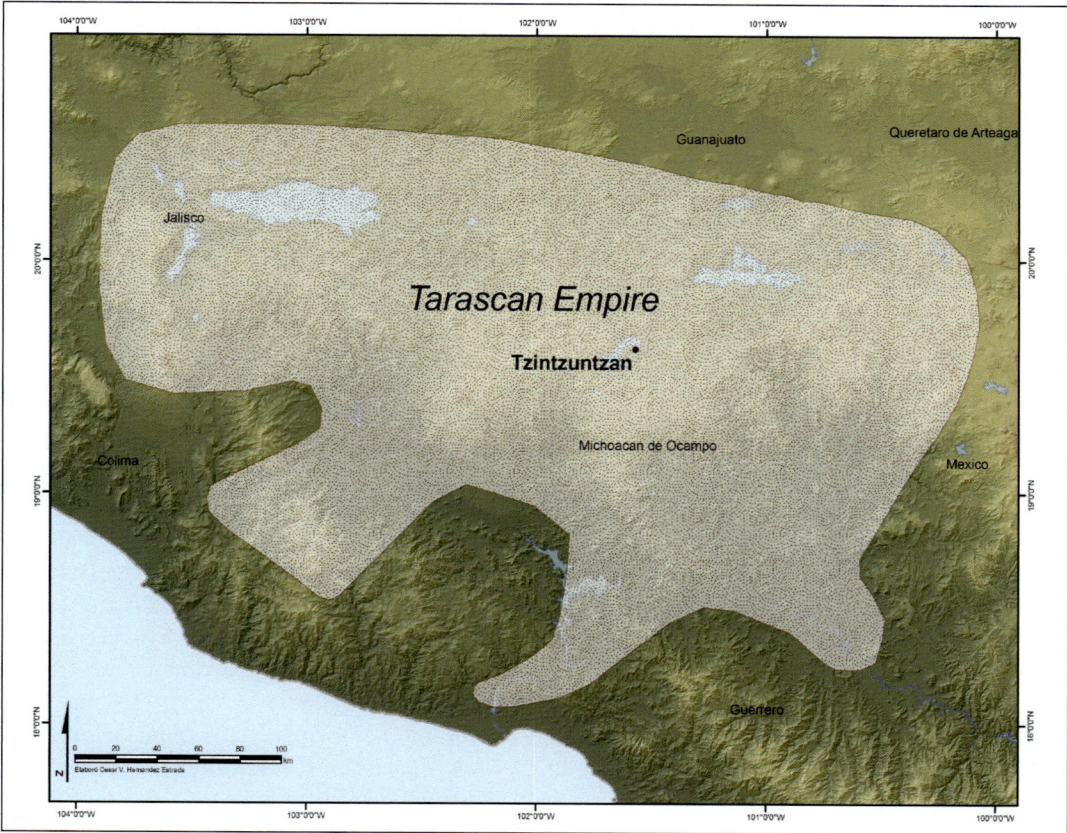

Figure 4.1 The Tarascan Empire in the framework of other Mesoamerican societies (Modified from Solanes and Vela 2000: 43).

Figure 4.2 Maximum extension of the Tarascan territory in 1522 (Adapted from Pollard 2000: Figures 5.1 and 6.2).

PERIOD	LOCAL PHASES	
LATE POSTCLASSIC	Tariacuri	(AD 1350-1525)
MIDDLE POSTCLASSIC	Urichu Tardío	(AD 1000/1100-1350)
EARLY POSTCLASSIC	Urichu Temprano	(AD 900-1000/1100)
EPICLASSIC	Lupe-La Joya	(AD 600/700-900)
MIDDLE CLASSIC	Jarácuaro	(AD 500-600/700)
EARLY CLASSIC	Loma Alta 3	(AD 350-500)
MIDDLE PRECLASSIC	Loma Alta 2	(100/50 BC-AD 350)

Figure 4.3 Chronology established for the Pátzcuaro Basin according to Pollard (2005: 9).

Basin) provided a direct means of increasing the access of the emerging elite to metal goods, primarily in the form of conquest booty and tribute. Metal adornments used as insignias of social status and public ritual became even more associated with political control. As part of a larger centralization of political and economic power by the royal dynasty, mining and metallurgy appears to have become to some extent a state industry by the sixteenth century. Pollard (1987: 748) points out that the zone of most active cultural assimilation of non-Tarascans to Tarascan ethnicity in the empire was the central Balsas region, at the major mining operations. She proposes that, by the sixteenth century, metal goods and the control of access to metal were no longer a reflection of individual social or ritual power, but of royal and state political power. Hence metal products were not just luxury goods used by the elite, but basic material expressions of political ideology and, therefore, essential to the maintenance of the prevailing social and political order.

Patterns of distribution and consumption of metal

Apparently, the use of fine metal goods within the Tarascan territory was restricted to the social and political elite. Implements used for subsistence activities, such as hoes and axes, were produced but not in quantity, and may even have been traded through local market networks within the Balsas Basin (Barrett 1987; Pollard 1987; Warren 1968). Other copper tools, including needles, awls, punches and fishhooks, appear to have been widely distributed, but their procurement and use was probably limited to craft or occupational specialists (Hosler 1994:156; Pollard 1987: 744). Presumably, all other metal goods were limited to the central dynasty and nobility (see Figure 3.8).

Excavations at Tzintzuntzan (Grinberg 1989; Pollard 1987; Rubín de la Borbolla 1944), Urichu (Hosler and Macfarlane 1996; Pollard 1995) and Huandacareo (Macias G. 1990), among other sites, have produced large numbers of metal objects from elite burials. This information is consistent with accounts in the *Relación de Michoacán* (Martínez S. 1903), probably one of the best known ethnohistorical sources on the Tarascans, which also refers to the existence of treasuries located at Tzintzuntzan and at six other centers in the Lake Pátzcuaro Basin. The metal stored in these treasuries was used exclusively in state religious ceremonies, under the custody of a head treasurer and other noblemen (Pollard 1987: 745).

Ore deposits and mining in the Tarascan territory

West Mexican metallurgy and metalworking, especially among the Tarascans and their neighbors, was based mainly on copper and its alloys (Hosler 1994: 127; Pollard 1987: 743). Most minerals that Tarascan metalsmiths processed abound in the territory of the Empire, including copper oxides and sulfides; native copper is also present. Artifactual evidence suggests that some or perhaps most of these sources of copper were being exploited from around AD 650 (see Hosler 1994). Prehispanic copper exploitation, however, reached its height during what Pollard (1982, 1987, 1993, 1997; see also Gorenstein and Pollard 1983) has called the Protohistoric period (AD 1450-1530) in the Tarascan domain. In the sixteenth century, the Tarascan language used the term *tiyamu charapeti haracuquaro* to refer specifically to copper miners (*tiyamu charapeti* meaning copper, or 'red metal') (see Warren 1991). Unfortunately, apart from linguistic facts and discrete references to several mines in Spanish Colonial documents (Grinberg 1995; Hosler 1994; Pollard 1987; Roskamp 2001, 2004; Warren 1968), little is known about copper mining in the Tarascan territory or in Mesoamerica in general.

Chapter 4: Tarascan copper smelting in the zone of Itziparátzico: a case study

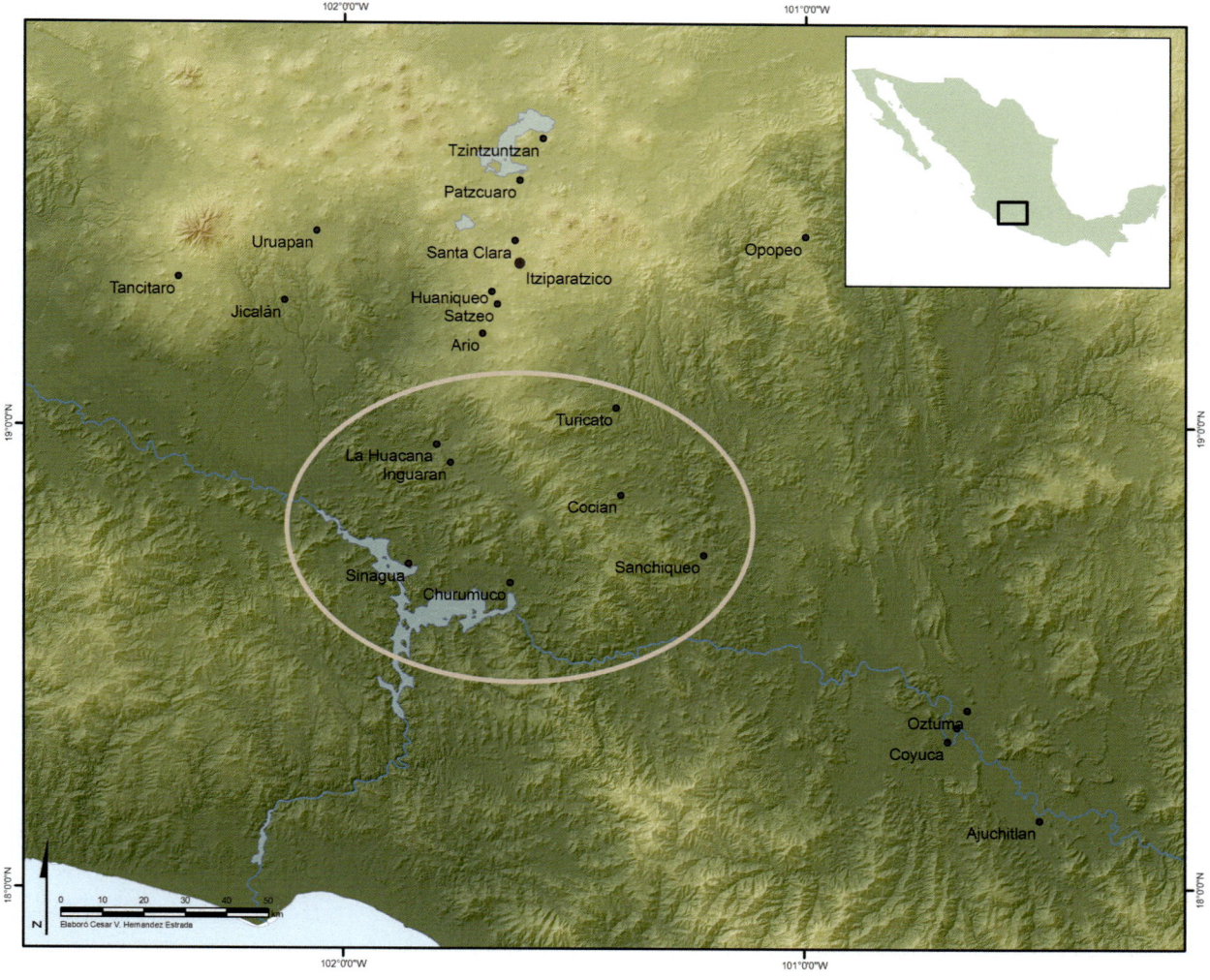

Figure 4.4 Mining centers in the Central Balsas Basin
(based on Barrett 1987: Map 2).

Hosler and Macfarlane (1996) carried out lead isotope studies of copper ores from several deposits in West Mexico, Oaxaca and Veracruz, as well as of a representative sample of copper artifacts recovered from excavations at a number of protohistoric sites in West Mexico and other areas in Mesoamerica. Among these sites was Urichu, one of the eight administrative centers of the Tarascan state. The results of these scientific analyses indicate that specific mining zones, Inguarán and Bastán (near La Huacana) in Michoacán, and the regions of Ayutla and Autlán in Jalisco, provided copper metal for Urichu and a number of other sites in West Mexico and beyond. Colonial documents (e.g. Martínez S. 1903) make extensive reference to Inguarán and Bastán and report copper mines in neighboring districts such as Sinagua and Guayameo, Tacámbaro and Turicato (Figure 4.4). The exact means of access to or procurement of the ore bearing resources, however, remain to be defined. More detailed data concerning mining and the organization of production in the Tarascan territory is presented in Chapter 5.

Smelting activities in the Tarascan territory

Although ethnohistorical sources provide some information on Tarascan extractive metallurgy, the technological processes that native metalworkers used to extract copper from its ores and the organization of the craft itself, remain poorly understood. Probably the most famous ethnohistorical source regarding Tarascan mining and metallurgy is the *Lienzo de Jicalán* (see Grinberg 1995; Roskamp 2001, 2004; see also Chapter 3 in this volume). This document illustrates metalworkers crouching in front of a brazier, heating metal by blowing through pipes (see Figure 3.4). Similar representations are shown in the *Relacion de Michoacán* (Martínez s. 1903: Plate XI) (see Figure 3.5) and in Central Mexican documents such as the Florentine Codex (Sahagún 1969-1982) and the Codex Mendoza (Berdan and Anawalt 1992) (see Figure 3.6). These illustrations, I believe, may represent melting of metal ingots for final processing, rather than the smelting of ores.

According to early accounts by Spaniards, the actual extractive operations were carried out by heating small quantities in shallow earthen pits, lined with a mixture of clay and ash, intensifying the heat by blowing through cane tubes (Barrett 1987: 15). Similar descriptions appear in other documents, including the *Legajo* 1204 (Warren 1968: 46, 48). Documentary sources also point out that with this method it was necessary to heat two or three times to produce metal of useful quality. I infer that part of this process may have involved roasting the ore to remove impurities before smelting.

After the spaniards took over the only significant technological change they introduced occurred some time after 1599, when blowpipes were replaced by animal skin bellows (see Barrett 1987). Evidence that little change took place is provided in a report to the Crown from officials in New Spain in 1574, in which they stated that copper production was largely in the hands of Indians and that because production was so low, it would not be worthwhile to tax it in the same manner as gold and silver (*Archivo General de la Nación*, [AGN] Minería, v. 30, exp 7, f. 154v, in Barrett 1987: 16). The excise tax (*alcabala*) was applied instead (Barrett 1987). According to testimony given in 1533 by Indians who worked in the mines, copper production was much lower than it had been before the Conquest (Bargalló 1955: 213, 295; Barrett 1987: 14). The low level of production was undoubtedly associated with the continuing decline in the indigenous population that followed the Spanish Conquest (Barrett 1987).

Artisans from Santa Clara del Cobre, a modern Tarascan municipality in Michoacán, have maintained the traditional techniques for working copper to this day. Essentially, the method employed by these contemporary craftsmen is based on the use of the *cendrada*, a hole in the ground lined with oak ashes (Figure 4.5a, b) which functions as a mold to produce a *tejo*, a disk-like copper ingot (Figure 4.5c) which is then hammered into finished shapes. Information provided by written sources, along with the factual presence of smelting waste products (i.e. slag) on the surface of archaeological areas near Santa Clara, suggest that the ongoing production of copper has its roots in prehispanic traditions (Horcasitas de Barros 1981; Maldonado 2002).

Santa Clara del Cobre is located approximately 14km south of Pátzcuaro and 29km from Tzintzuntzan (Figure 4.6). The community was officially founded when a curacy (which also included the neighboring town of Opopeo) was instituted in 1577, although it had started as a congregation settlement twenty years earlier (Paredes 2004). Presumably because of its strategic location in the pine-oak uplands, where forests would provide an ample supply of charcoal (and there would be enough charcoal makers), in 1604

Figure 4.5a *Cendrada* lined with oak ashes, used by contemporary coppersmiths from Santa Clara del Cobre (Photo by Patricia Castro, 2018).

Figure 4.5b Coppersmith from Santa Clara del Cobre melting scrap copper.

Figure 4.5c Resulting red-hot copper *tejo* (ingot).

Chapter 4: Tarascan copper smelting in the zone of Itziparátzico: a case study

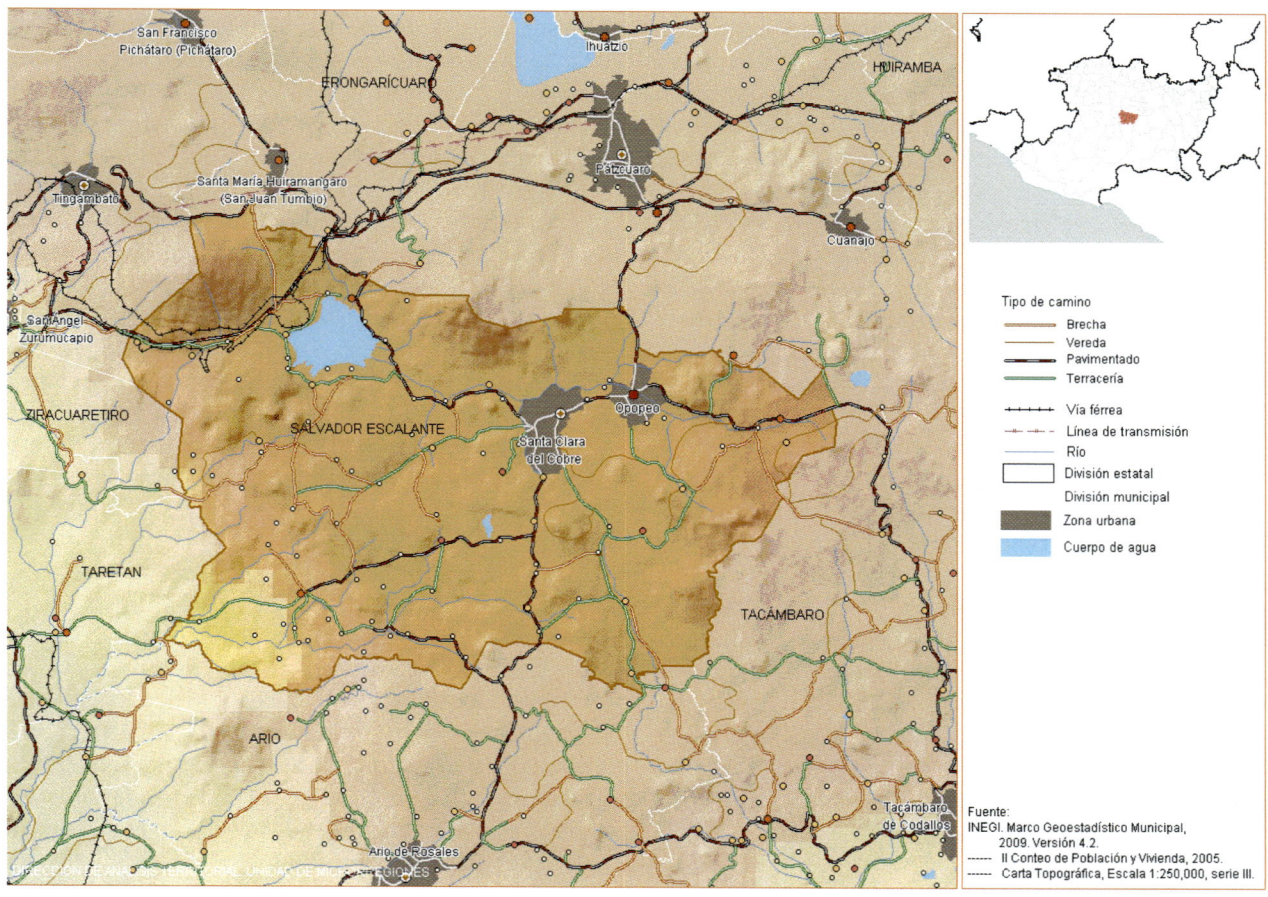

Figure 4.6 Santa Clara del Cobre-Opopeo
(Based on unless otherwise stated, all courtesy of SEDESOL, Gobierno Federal: http://www.microrregiones.gob.mx/zap/datGenerales.aspx?entra=zap&ent=16&mun=079.

Santa Clara del Cobre was chosen to establish the royal copper smelter under the administration of Spanish officials. The community of Santa Clara would become the most important smelting center in New Spain, in spite of its location about 125km away from the mines that supplied the ore, in the region of La Huacana (Barrett 1987; Moreno 1986; Paredes 2004). The Spanish Bishop Vasco de Quiroga (1533, in Warren 1968) pointed out that since the smelting of ore took three times as much charcoal as ore, it was probably more economical to transport the ore from the mines to the source of charcoal than the other way around.

Documentary sources are not clear as to whether metalworkers in Santa Clara were local, or had been congregated from the mining region. One particular account referring to the difficulties that the Spanish had in obtaining enough workers due to the dramatic decline in native population during the sixteenth century, states that in 1604 the viceroy granted the right to use smelters from Opopeo to a copper administrator because 'they were the only ones left who knew this type of work' (*Archivo General de la Nación* [AGN], *Reales Cédulas Duplicadas*, v. 16, exp 346, f. 108, in Barrett 1987: 23). This strongly suggests the existence of metalworkers in the area prior to the Spanish arrival. It also provides an additional reason for Spaniards to locate their smelting operations there. Recent archaeological investigations at the archaeological zone of Itziparátzico (see Figure 1.1), have documented new evidence for studying aspects of pre-Contact extractive metallurgy of copper in the Tarascan domain.

Physiographic background: the Santa Clara-Opopeo region

The Santa Clara-Opopeo region is located in the Zirahuén Basin, within 19°24'45' north latitude and 101°38'45' west longitude of the Greenwich Meridian, at an altitude of 2,298m above sea level (a.s.l.) (Guevara F. 2004; Morán Z. 1984). It covers an area of roughly 460sq km and borders Pázcuaro to the north, Ario de Rosales to the south, Huiramba and Tacámbaro to the east, and Ziracuaretiro, Tingambato and Taretan to the west. Santa Clara del Cobre is the *cabecera* of a *municipio* that is divided into two *tenencias*: Opopeo and Zirahuén (Guevara F. 2004).

The region lies on the southern boundary of a large volcanic belt stretching across central Mexico from 18°- 22°N and often referred to as the Trans Mexican Volcanic Belt or Neovolcanic Axis, which is characterized by recent (Pliocene-Quaternary) volcanic activity involving numerous monogenetic (one eruptive phase) cinder cones and shield volcanoes. These conditions have resulted in a densely lineated and ridged terrain. The topography of the Santa Clara-Opopeo region is dominated by the Sierra of Santa Clara, a number of isolated hills and some small intermontane plains (Ferrari et al. 1994; Hasenaka and Carmichael 1985).

The area with which this study is concerned is surrounded by alluvial plains and low mountainous areas to the north and south, by the Frijol, El Burro, Zimbicho, La Tapada and Cofradía hills (about 3,000 m a.s.l.) to the northeast, and by Cerro Zirahuén (3,000 m a.s.l.) to the northwest (Guevara F. 2004: 68). The Santa Clara region is part of an Upper Cenozoic volcanic belt consisting mainly of rocks dating from the Late to the Recent Pliocene Period, with the predominant types being lava stone and basalt and volcaniclastic basalt breccias appearing generally at the summit of volcano cones and on the flanks of certain mountains (Ferrari et al. 1994: 97; Guevara F. 2004: 68).

Hydrology and climate

In terms of hydrology, the Santa Clara-Opopeo region lies on the south-central part of the Lake Zirahúen Basin, which is bounded to the north by the Lake Pátzcuaro Basin. The boundary between these two watersheds is marked by the Sierra of Santa Clara, including the Zirahuén, La Cantera, El Tecolote and El Frijol hills. To the northwest, the hill El Burro separates the Zirahuén Basin from the Lake Cuitzeo Basin. To the east and southwest, the Zimbicho, Janamo, La Tapada and La Cofradía hills stand between the Zirahuén Basin and the Tacámbaro Basin. To the south, the Zirahuén Basin is bounded by a series of ridges descending like a natural staircase from the highlands to the lowlands of the Balsas River Basin (Guevara F. 2004; Pérez-Cálix 1996:73).

The hydrography of Santa Clara-Opopeo is characterized by water inflow from the rivers El Silencio (also known as Río de la Palma) and Los Manzanillos, the Turirán and Agua Blanca streams, the lakes of Zirahuén and Cuizitan, and various cold-water springs. There are no major rivers in the region. The largest aquifer in the study area is El Silencio River, which flows from east to west and empties into Lake Zirahuén. El Silencio is fed mostly by ephemeral streams and a few intermittent watercourses that flow from the uplands. The main springs in the area are San Gregorio, Cungo, Las Palmitas, San Miguel, Cherahuén, and Santa Clara del Cobre (Guevara F. 2004: 69).

The Santa Clara-Opopeo region has a Type C (w2)(w) climate, which corresponds to a sub-humid temperate kind with summer rains (Guevara F. 2005). Average annual precipitation is 1182 mm yr^{-1}. Most precipitation (95%) falls in summer, between June and September, July being the wettest month. Less than 5% of annual rainfall occurs in the winter. The dry season is from December to April. Average annual temperature is 15.7°C (60°F) (Davies 2005; Guevara F. 2004). It is higher in the lower areas to the south and up to 10°C (18°F) cooler in the mountains. The record low temperature in the lowlands is 3°C (38°F), and temperatures can be as low as 0°C (32°F) in the uplands. The record high temperature is around 27°C (80°F) at lower altitudes and 18°C (65°F) at higher elevations (Guevara F. 2004: 70).

Soils and vegetation

The soils in the study region are almost exclusively volcanic types derived from ash and extrusive igneous rock, although alluvial soils also occur. Both types were formed relatively recently (Alcalá de Jesús et al. 2001). The main soil types found include andosol, litosol, luvisol and phaeozem. The regional native name for andosol is *tupure*. It has a sandy texture and ranges from brown to black. The organic matter content is high (4 to 6%), which is a typical feature of humic andosols. Both structure and texture allow excellent drainage with good moisture and nutrient holding capacity, which makes it highly valued for agriculture (Alcalá de Jesús et al. 2001; Etchevers B. 1985).

Litosol is known locally as *malpais* and consists of very thin soils that vary in depth from 0-50cm. The texture is either loamy or sandy; organic matter content is high with intermediate content of nutrients. These shallow soils are found overlying rocky hills and ridges (Alcalá de Jesús et al. 2001). Luvisols have a red color with a clay texture and are found mainly in the upland areas. Local farmers recognize three subtypes on the basis of textural and color differences: *jaboncillo* has a homogenous, doughy texture and bright red color, and is used to finish floors and flower boxes; *barro rojo* (red clay) has a coarser texture than *jaboncillo* and is used to make bricks and tiles and as mortar for fireplaces and chimneys; *charanda* has an even coarser texture than clay and an opaque red color, and is used mainly to grow fruit (Alcalá de Jesús et al. 2001; Guevara F. 2004).

Finally, phaeozems are found in alluvial plains and characterized by a humus-rich surface layer covered in the natural state with abundant grass or deciduous forest vegetation. These are highly arable soils used for growing crops, as well as for wood and fuel production (Alcalá de Jesús et al. 2001). Local farmers in the region also recognize *tepetate* as a soil type. Although the word is used to refer to a wide range of rocks and soils in the literature and by Mexican farmers, Williams (1972) suggests that it is most often equated with caliche, and is a hard, limey duricrust. Although *tepetate* is often found as a distinct hardpan layer underlying the soil, there are some areas where erosion has removed most or all of the covering soil layer, and farmers have brought the *tepetate* into cultivation. In the Santa Clara-Opopeo region, this sandy milky-yellow soil is occasionally mixed with *tupure* to grow corn.

Vegetation is probably the environmental aspect most directly relevant to the present work, since it provided the fuel required for the smelting processes carried out at Itziparátzico. The vegetation of Santa Clara-Opopeo falls into the Temperate Vegetation Types Zone (Leopold 1950). Areas of evergreen boreal forest, and pine-oak forest and mixed forests still remain in the region, as well as grasslands and riparian, aquatic and sub-aquatic flora. Sacred fir (*Abies religiosa*) forest is found in high elevations at altitudes ranging from 2,800-3,300 m a.s.l., normally in association with nutrient rich andosols in protected areas.

Sacred fir forests often form pure or nearly pure growths, but can also be found intermixed with pine and oak, including *Pinus Michoacána, Pinus pseudostrobus, Quercus laurina*, and *Quercus rugosa*. Pine forest appears at elevations of 2,100-3,000 m a.s.l., associated with well-drained andosols overlying extrusive volcanic basalt and breccias. The most widely distributed species is *Pinus pseudostrobus*, but *Pinus Michoacána* and *Pinus leiophylla* are also common (Guevara F. 2004; Leopold 1950; Pérez-Cálix 1996). Pine forests in the region are currently being overexploited and the surviving forests are increasingly restricted to relatively inaccessible locations.

Oak forest tends to be found in foothills at altitudes of 2,000-2,600 m a.s.l. throughout the Zirahuén Basin. This vegetation type borders the lower limits of the pine forests and often appears on rough topography such as the extrusive volcanic rock outcrops common in the area. Oak forests occur in association with three soil types: andosols, phaeozems and litosols. Although oak trees tend to be deciduous, they give the appearance of being evergreen because the leaves grow back so quickly. At higher altitudes the dominant species is *Quercus rugosa*, while *Quercus castanea* is more frequent in lower elevations. The herbaceous layer is highly diverse and includes a large number of species belonging to the *Compositae, Leguminosae*, and *Graminae* families. A variety of epiphytes including orchids and ferns is also found. Arboreal communities between 2,100-2,300 m a.s.l. consist of montane mesophyll forests or cloud forests, which are commonly found on mountain sides, ravines and isolated areas protected from the wind and direct sunlight. They grow on moderate to steep slopes of igneous basalt substrate where the dominant soils are andosols. The most common species include mountain alder (*Alnus acuminate*) and bonpland willow (*Salix bonplandiana*), as well as several types of pine and oak (Guevara F. 2004; Leopold 1950).

Grasslands are characterized as areas dominated by grasses (Poaceae) rather than large shrubs or trees. Two types of grasslands are found in the Santa Clara-Opopeo region; montane grasslands and lake-shore grasslands. The former are found at high altitudes (3,300+ m a.s.l.). The dominant species is *Muhlenbergia gigantea*, which appears in dense colonies about 1 meter high and grows alongside other grasses such as *Agrostis tolucencis, Bromus carinatus* and *Festuca amplissima*. The other type of grassland grows mainly on the shores of Lagunita de San Gregorio, at about 2,700 m a.s.l., as a carpet of short (40-60cm) perennial grasses. These tend to be widely scattered without forming dense colonies. The most common species include *Aegopogon cenchroides, Agrostis schaffneri* and *Bromus carianatus*, among others. These are often found in association with a number of herbaceous and undershrub species. There are also some induced grasses that have grown in the aftermath of the devastation of various types of vegetation. Riparian vegetation consisting mainly of 5-10 m tall trees is found on the western shore of Lake Zirahuén and on streambeds. The main species include *Alnus acuminata* (mountain alder) and *Salix bonpladiana* (bonpland willow) (Guevara F. 2004; Leopold 1950).

Aquatic and subaquatic vegetation is found mainly in the area of Lake Zirahuén, which hosts numerous submersed hydrophytes (e.g. *Potamogeton foliosus* and *Potamogeton illinoensis*), emergent hydrophytes (e.g. *Cyperus canus, Cyperus hermafroditus, Juncus ebracteatus*, and *Scirpus californicus*) and small amphibious hydrophytes (e.g. *Aster subulatus, Berula erecta, Bidens aurea, Sagittaria latifolia* and *Sagittaria macrophylla*). Hydrophytes also grow in seasonally or permanently flooded/ponded areas with soils that remain semi-permanently or permanently saturated. This can be observed in the immediate surroundings of Santa Clara del Cobre and Opopeo, where the El Silencio River forms still pools along its course. The herbaceous species found in this area (some growing next to the ground, others reaching up to 1 m tall) include *Aster subulatus, Berula erecta, Bidens aurea, Bidens odorata*, and *Eriochloa holciformis* (Guevara F. 2004).

History of land use

The lake Zirahuén Basin has an auspicious geographic location at the juncture of the Tarascan core area and the main route leading from central Michoacán to the Balsas Basin in the *Tierra Caliente*. The richness of natural resources enhanced the appeal of this region for human settlement. Important natural features of the region include:

1. Its location in an area of intermontane valleys

2. A climate gradient of good rainfall and favorable temperatures resulting in relatively high humidity
3. A river and several springs, which ensure a reliable water supply; proximity to Lake Zirahuén and its associated aquatic resources.

In spite of these advantages, archival evidence suggests that settlement and exploitation in the district of Lake Zirahuén had been apparently negligible even up to the turn of the seventeenth century. According to ethnohistoric sources (e.g. *Archivo General de la Nación* [AGN], *Tierras*, 3127, exp. 1 fs. 30 and 35, in Endfield and O'Hara 1999: 412), land around the lake was reserved largely for ceremonial and recreational purposes by the Tarascan nobility. Prior to this study, no archaeological work had been carried out in the basin, and hardly any references to prehispanic settlements have been found in early Colonial documents, unlike other lake basins in the region, notably Pátzcuaro and Cuitzeo. This suggests that the basin was not densely occupied (Endfield and O'Hara 1999). A detailed archaeological survey throughout the basin would be required to verify this. This information is relevant because it seems to place prehispanic copper metallurgy at Itziparátzico in an elite-controlled context (further discussion of this matter is provided in Chapter 6).

Very little is known about the Zirahuén Basin in the early Colonial Period either. Few references are found in historical documents, which may be significant, given the great importance that the Spanish placed on keeping detailed written records. One rare description from 1619 describes the basin slopes as forested (*Archivo General de la Nación* [AGN], *Tierras*, 3695 *Expediente* 7, in Davies et al. 2004). The area simply may not have been attractive to Spanish settlers since they sought to settle prehispanic loci where they could take advantage of existing indigenous systems of food production, well-established trading and infrastructural networks and, perhaps more significantly, dense tribute-paying populations (Endfield and O'Hara 1999; Paredes 1984). The eighteenth century, however, saw a marked increase in Hispanic settlement and exploitation in the Zirahuén Basin. By the mid-1700s, a number of large private estates, known as *haciendas*, were established to grow sugar, maize and wheat and to raise cattle (Endfield and O'Hara 1999).

Some documents indicate environmental degradation in the basin around this time. By 1733, land in the southern part of the basin had been 'stripped of vegetation' (AGN *Tierras* 514 *Expediente* 3, in Davies et al. 2004), while deforestation to the southwest of the lake had led to gullying (AGN *Historia* 73, *Fojas* 334-335, in Davies et al. 2004). It appears that the steeper high ground surrounding the basin may have escaped deforestation. In 1789, Lake Zirahuén was described as 'surrounded by high hills of pine' (AGN *Historia* 73, *Fojas* 391-392, in Davies et al. 2004) and the view around Santa Clara-Opopeo was of 'mountains and hills populated with pine and some oak' (AGN *Historia* 73, *Fojas* 391-392, in Davies et al. 2004).

The most important development in the Zirahuén Basin during the Colonial period was the intensification of copper smelting works in the town of Santa Clara del Cobre. Santa Clara became the principal copper smelting centre in Mexico, and by 1789 there were eight royal smelters, each employing between 30 and 40 workers (AGN *Historia* 73, *Fojas* 391-392, in Davies et al. 2004). By the turn of the 20th century, the copper industry at Santa Clara had diminished greatly. Today, only a handful of coppersmiths remain, producing goods from scrap copper. The pattern of development in the Zirahuén Basin during the Colonial Period is illustrated by the population records from Santa Clara. In 1600, there were just 130 inhabitants (Carrillo-Cázares 1996), but by 1822, the population had risen to 2813 (Martínez de Lejarza 1974).

The two largest modern towns in the basin are Santa Clara and Opopeo, with a combined population of around 20,000 (Santa Clara being the most important one, with a population of about 12,000) (INEGI 2000). The major agricultural activity in the basin is maize cultivation, although small amounts of vegetables and wheat are also produced. There have been, however, considerable changes in land use throughout the basin in recent years. Agricultural activities intensified significantly with the development of a commercial fruit farming operation on the south lakeshore (Davies et al. 2004). Native forest species are still exploited for wood and lumber. Certain pine varieties (particularly *Pinus leiophylla*, *Pinus Michoacána* and *Pinus pseudostrobus*) also provide resin (Guevara F. 2004). These activities, combined with deforestation to provide charcoal for copper smelting, have caused severe soil erosion and environmental degradation in the area over the last couple of hundred years (Davies et al. 2005).

Itziparátzico: the research project

The archaeological zone of Itziparátzico is located in the south-central part of the Lake Zirahúen Basin in north central Michoacán, lying among the modern agricultural fields of Santa Clara del Cobre and Opopeo (see Figure 1.1; also Figure 4.7a, b). Although previous archaeological investigations in the region was concentrated in the vicinity of Lake Pátzcuaro particularly in the Tarascan capital of Tzintzuntzan, it is well known among the older coppersmiths of Santa Clara that numerous mounds or *yácatas* lie buried in this area. It is important to point out, however, that while the word *yácata* is used locally to refer to any man-made mound, in archaeological literature the

Chapter 4: Tarascan copper smelting in the zone of Itziparátzico: a case study

Figure 4.7a View of Itziparátzico
(Photo by J. Silverstein, 2004).

Figure 4.7b View of Itziparátzico from the terraces.

word is limited to the keyhole-shaped structures associated with Tarascan state religion, which have not been identified at Itziparátzico.

The main goal of this research was to explore evidence for prehispanic metallurgy at Itziparátzico. The project was designed to achieve four major objectives:

1. To locate, surface survey, and map all evidence for prehispanic metalworking at Itziparátzico.
2. To undertake preliminary archaeological test excavations in production areas to confirm the existence of in situ craft activity and obtain stratigraphic samples of material remains to date metallurgical production deposits.
3. To analyze archaeological samples recovered from excavation for microstructure and composition.
4. To use the results of these investigations to begin to build a model of Tarascan metallurgical production in prehispanic West Mexico.

The purpose of this study, then, was to identify traits, or clusters of traits, associated with different stages of prehispanic metallurgy. The results would provide the basis for comparing archaeological evidence for metallurgical production with descriptions of prehispanic metallurgy found in ethnohistorical sources. These comparisons were intended to examine both the technology and organization of production. This work represents the first fully reported study of copper smelting in Mesoamerica, as well as the first systematic archaeological investigation carried out in the Zirahuén Basin.

Preliminary work

As with any archaeological project of this nature the preliminary step to fieldwork was ethnohistorical background research designed to provide as much information about the area of study as possible. My research involved examining several documents, including the various *Relaciones Geográficas* of 1579-1580, the *Relación de Michoacán* (1541) and the *Lienzo de Jicalán* for references to Tarascan towns and villages where metal production took place. During the summers of 2000 and 2002, I carried out an ethnoarchaeological study of contemporary metalworking in Santa Clara del Cobre.

My research included working with modern coppersmiths to conduct experiments with native copperworking technology (Maldonado 2001, 2002). I interned at the CECATI in Santa Clara during the summer of 2002, where I learned metallurgical processes still employed by local copper craftsmen. I replicated Tarascan metal axes, tweezers and bells in order to estimate the prehispanic energetic inputs in the manufacture of the original artifacts (see Maldonado 2005). Combined ethnohistorical and ethnoarchaelogical information was useful for framing research questions to be answered with a field investigation and to provide context for the findings, as well as insights about the people who once lived in the area now known as Itziparátzico.

Surface survey

Located by the El Silencio River, the archaeological zone known locally as Itziparátzico is about 1.5km NW of Santa Clara del Cobre and the same approximate distance from Opopeo (see Figure 1.1 and Figure 4.8). Itziparátzico lies on the transition between alluvial plains and low mountainous areas of pine and oak forests (Figure 4.7a, b). The U.T.M. coordinates are 2,158,390 North and 219, 820 East (Topographic Chart E14A32 scale 1:50,000, INEGI 2001). Preliminary surface survey during the summer of 2002 confirmed the presence of prehispanic mounds, domestic terraces and evidence of metalworking. Metallurgical evidence consisted largely of slag (known locally as *querenda*) found in discrete concentrations across the area.

Previously, Cárdenas G. (1986) had reported the existence of three archaeological sites in this zone, which he identified as Potrero La Cornejalera, Puente del Rebocero and Opopeo. This report, however, was largely based on partial reconnaissance data. With the support of FAMSI and Penn State University, I initiated systematic archaeological investigations at Itziparátzico in the summer of 2003. During the Itziparátzico Archaeological Research Project (IARP) 2003-4, approximately five weeks were dedicated to surveying and mapping natural and cultural features in the zone described by Cárdenas and its surroundings. While Itziparátzico may actually represent more than one settlement, no clear site limits were identified during this investigation.

The principal goals of the archaeological survey were to define the extent and nature of the zone of Itziparátzico, to identify and characterize its chronology, and to locate and record evidence for metallurgical activity. Natural features in the area, including terraces and springs, as well as cultural features such as mounds, roads and agricultural fields were mapped using *Global Positioning System* (GPS) instrumentation in combination with a digital topographic chart (scale 1:50,000, INEGI 2001), aerial photographs (scale 1:25,000, INEGI) and digital orthophoto images (scale 1:75,000, INEGI 1995).

This equipment was employed to map an area of approximately 15sq km and record archaeological findings. The GPS also served as an excellent reconnaissance tool that could be used to create spatially accurate sketch maps. Systematic surface

Chapter 4: Tarascan copper smelting in the zone of Itziparátzico: a case study

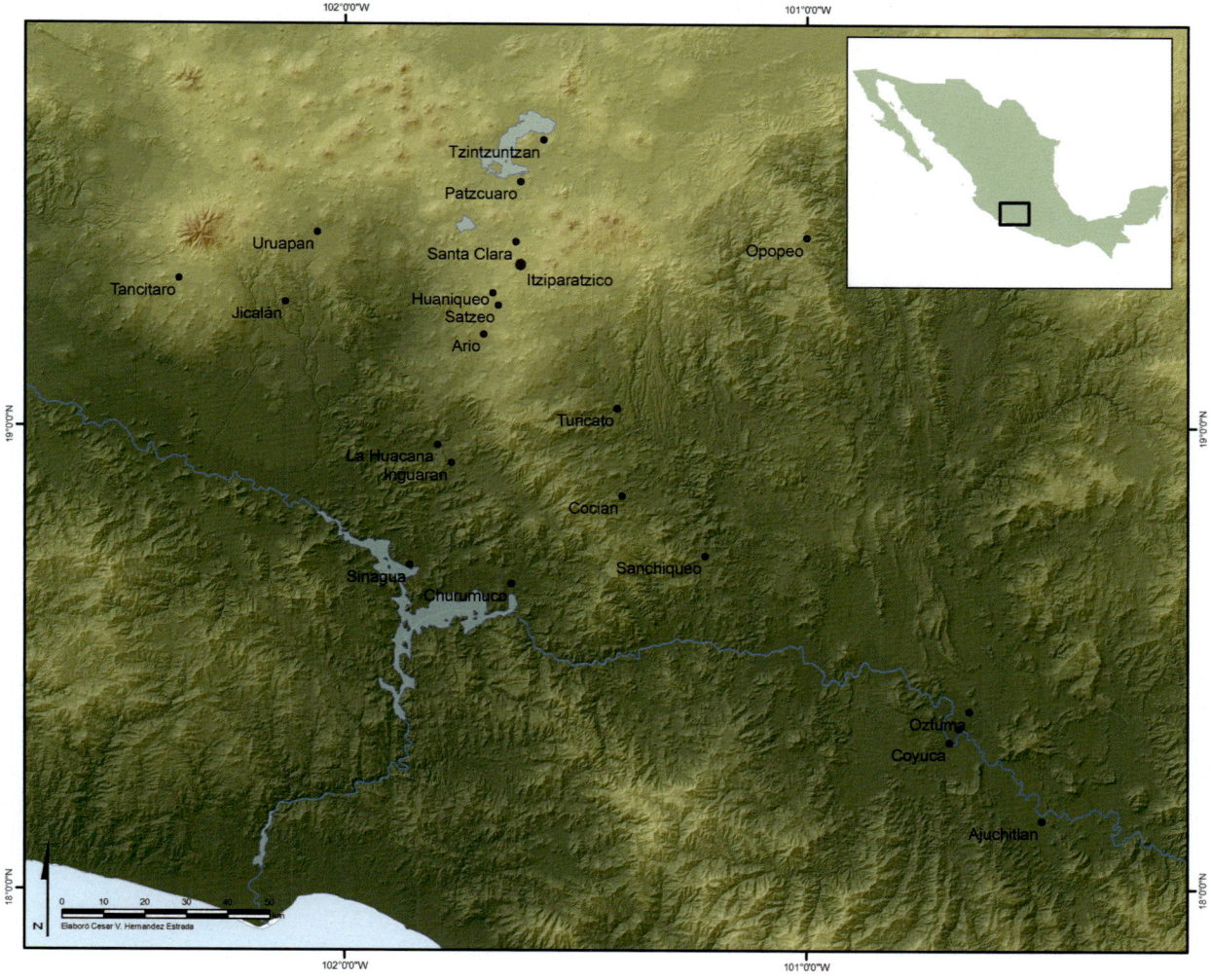

Figure 4.8 Location of Itziparátzico and other mining and smelting localities within the Tarascan territory (Map by César Valentín Hernández, based Roskamp et al. 2003: Figure 4).

survey was carried out along 29 agricultural fields, using parcel boundary lines as reference points rather than following a grid system. A full-coverage survey aimed at obtaining a wide record of the spatial variability in the area (and do a proper selection of sites for excavation) was also conducted.

Intensive surface survey was used to locate production areas represented by concentrations of manufacturing byproducts (i.e. slag). Archaeological materials are present at varying densities throughout the site. A total of 79 surface collections were made during the survey including ceramics, lithics (mainly obsidian) and slag. The surface survey was performed by a team of 2-3 archaeologists and 3 local assistants walking the fields at an interval of 5-10 m between them, depending on the shape of each field and surface conditions.

The team walked in line across fields while continually communicating relevant information for mapping and recording on forms. Multiple passes were required for large fields. Concentrations were recorded on the 1:25,000 aerial photograph and later mapped in on the digital map using GIS (Arcview). For each concentration, at least one 3m radius circular 'dog leash' collection was made of 100% of all artifacts and slag within the circle.

Throughout the parcel survey, isolated artifacts were also collected, on a regular basis, whether or not they were within a concentration. We collected isolated artifacts if they represented highly diagnostic formal tool types of varieties not represented in our systematic collections (or represented by only a few examples). The location of isolated collections was recorded relative to the nearest structure. The survey recorded a total of thirty-one earthen structures labeled as 'mounds'. They vary in size and shape, ranging from 1-meter high circular structures to 10-meter high quadrangular constructions of earth and stone. Five of these mounds appear to be culturally-modified natural elevations. All

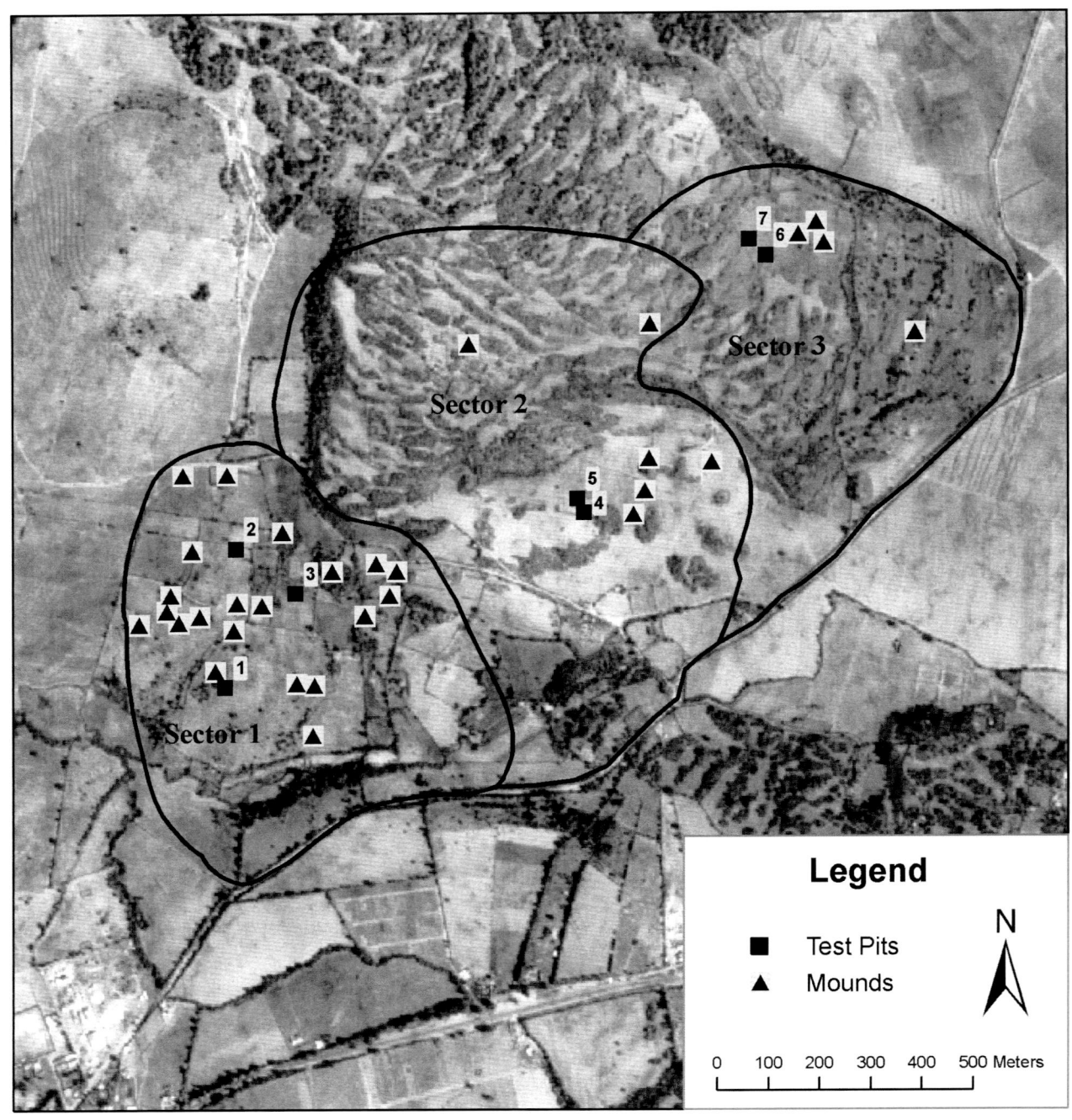

Figure 4.9 Three main sectors of the Itziparátzico area
(Map created by P. van Rossum, 2006).

of these structures have suffered from a certain degree of damage due to farming activities in the area. Most of them show clear evidence of looting (see Appendix A for detailed descriptions of the structures).

Surface survey identified and mapped three major sectors of the site, divided according to the variability of their archaeological materials and features (see Figure 4.9). While ceramic and lithic artifacts were common throughout the research area, slag concentrations were located almost exclusively in one particular sector (Sector 1). Both the presence of smelting byproducts and the proximity to water (indispensable for metalworking processes) indicate that smelting activities may have taken place in or around this zone. Other materials include moderate amounts of potsherds and lithics (mainly gray-black obsidian prismatic blades), as well as a set of stylistically diverse Tarascan pipes. Pollard (personal communication) has positively identified these pipes and several polychrome ceramic fragments as Late Postclassic in date (see Figure 4.3), and suggested that the source of the gray-black obsidian may be

Ucareo. Out of the thirty-one earthen mounds recorded at Itziparátzico nineteen were located in Sector 1 (Map 4.6). General descriptions of the materials recovered from surface at Iztiparátzico are provided in Appendix A.

The second sector of the survey area (Sector 2) consists basically of terraces, both occupational and agricultural. A total of twenty-seven terraces, most of them domestic in nature, were identified in this sector. Five mounds were also recorded in sector 2 (Figure 4.9; see also Figure A.2 in Appendix A). Although some slag is scattered on surface, the amounts are significantly smaller than those at Sector 1. On the other hand, Sector 2 presents the highest densities of potsherds and lithics on surface. Some of these potsherds present painted decoration. Lithics include gray-black obsidian blades and a few grinding stone fragments. Two anthropomorphic figurine fragments were also collected from one of the terraces.

The third sector (Sector 3) contains the five largest mounds recorded in the survey area, including Mound 25, a square-shaped structure of about 40 x 40m and 10m high, built of earth and stone (Figure 4.9; see also Appendix A). The density of materials on surface, however, seems lower in this sector than in Sector 1 and 2. Both the nature of the archaeological artifacts throughout the zone of Itziparátzico and their density on surface (0-20 specimens/m^2) seem to indicate that the area experienced its most significant occupation during the Late Postclassic Period (see Figure 4.3). The patchiness of the materials on surface also suggests a scattered settlement pattern. The results of this surface survey set the parameters for the selection of excavation areas.

Test pitting

Sample collection from stratigraphic contexts was the next imperative step of my research, which was achieved through nine weeks of intensive fieldwork beginning in mid January and ending in late March 2004. This fieldwork involved the preliminary archaeological test excavations in the three major sectors of Itziparátzico: the sector near the water springs, which presumably represents the metal production zone of the site; the sector of the domestic terraces and the sector of the large mounds (see Figure 4.9). Seven test-pits were excavated through deposits of silt and clay in different parts of the research area. The overall aim was to assess the quality and significance of the archaeological information obtained during the surface survey. Three 2 x 2m test pits were located in Sector 1, two in Sector 2, and two 2 x 1m units in Sector 3 (Figure 4.9; see also Appendix B). All metallurgical remains, pottery, lithics, soil samples, and soil and radiocarbon samples were collected. The results of these excavations were consistent with the observations on surface. No substantial evidence for occupation before or after the Late Postclassic Period was found.

All of the ceramic and lithic artifacts from excavation seem consistent with a Late Postclassic Tarascan occupation. The lithics include gray-black obsidian blades, arrowheads, modified flakes, and cores (Figures 4.10 and 4.11). Basalt blades, axes and hammers were also recovered, as well as grinding stones (Figures 4.12 and 4.13). Discrete amounts of potsherds were recovered including both undecorated and decorated samples. The most abundant fragments correspond to a domestic, well-polished, red-slipped ware (Figure 4.14). Although the general condition of the pottery is highly fragmentary, we have been able to identify pieces of jars and bowls, as well as two fragments of stirrup-spouts (see Figure 4.15). The most common type among the decorated specimens is a red-and-white-on-cream ware, which seems to have included forms such as bowls and plates (Figure 4.16, Figure 4.17). Other polychrome examples in our collection are shown in Figures 4.18, 4.19, 4.20 and Figure 4.21. Incised and appliqué decoration is also represented in the sample (Figures 4.22 and 4.23). The ceramic assemblage also includes a number of Tarascan pipe fragments (Figures 4.24 and 4.25). Detailed descriptions of the materials recovered from stratigraphic contexts are provided in Appendix B.

Regrettably, no identifiable metalworking structures (furnaces, hearths or pits) were found at Itziparátzico during the test-pitting. Slag samples recovered from excavation therefore represent the most relevant material for the purposes of this project. While these smelting waste-products were recovered in large amounts, only a representative sample of 2.1kg was selected from several hundreds of kilos and exported for metallographic analysis. Part of this material has been analyzed at Wolfson Archaeological Science Laboratory, Institute of Archaeology, University College London. The absence of metallurgical materials other than slag (i.e. fuel, hearth structures, crucible fragments, mould fragments, stock metal, metal prills, failed castings, part-manufactured objects and spillages, etc.) around Itziparátzico appears to indicate that only primary copper production was being carried out at the site. Primary production involves the actual smelting or extraction of metal from its ore by heating. Conversely, secondary production or smithing implies the working or forging of metals into artifacts (Bachmann 1982) and produces a diagnostic range of remains, such as mould and crucible fragments, and metallic debris.

Summary description of Itziparátzico

The archaeological zone of Itziparátzico is located in the Zirahuén Basin, in a transitional area between small

Tarascan Copper Metallurgy

Figure 4.10 Lithics from Itzparátzico: Gray-black obsidian arrow head, modified flake and prismatic blade.

Figure 4.11 Lithics from Itzparátzico: Gray-black obsidian core.

Figure 4.12 Lithics from Itziparátzico: Basalt tools.

Figure 4.13 Lithics from Itziparátzico: Basalt axe.

CHAPTER 4: TARASCAN COPPER SMELTING IN THE ZONE OF ITZIPARÁTZICO: A CASE STUDY

Figure 4.14 Polished red slip ware from Itziparátzico.

Figure 4.15 Red slipped stirrup spouted vessel fragment.

Figure 4.16 Red-and-White-on-Cream rim and body fragment.

Figure 4.17 Red-and-White-on-Cream body fragment.

Figure 4.18 Red ware with resist (negative) decoration.

Figure 4.19 Decorated Red-and-Black-on-White pottery.

Figure 4.20 Decorated Black on White pottery.

Figure 4.21 Decorated White on Red pottery.

Figure 4.22 Potsherd showing incised decoration.

Figure 4.23 Potsherd showing appliqué decoration.

Figure 4.24 Zoomorphic pipe fragment.

Figure 4.25 Pipe fragment showing incised decoration.

intermontane valleys and the pine-oak uplands. Three archaeological sites had been reported previously in this zone (Potrero La Cornejalera, Puente del Rebocero and Opopeo). That report, however, was largely based on partial reconnaissance data and the limits of the sites were not established at the time. While Itziparátzico may indeed represent more than one settlement, the IARP 2003-4 mapped an area of approximately 15sq km and found no clear site delimitations.

Surface survey identified and mapped three major sectors of the site, divided according to the nature and density of their archaeological materials and features. Ceramic and lithic artifacts were common throughout the research area. Smelting byproducts (i.e. slag), however, were located almost exclusively in Sector 1. The presence of slag indicates that smelting activities may have taken place in or around this area. Sector 2 of the zone consists basically of domestic terraces (twenty-seven in total), while Sector 3 contains five of the thirty-one earthen mounds recorded in the area, including a 40 x 40m and 10m high structure.

Ceramic evidence recorded during the IARP 2003-4 surface survey suggested that the settlement had only a single occupation or component, dating to the Late Postclassic period (see Table 4.1). The concentration of the materials on surface suggested a scattered settlement pattern. Test-pitting operations carried out in the three sectors of Itziparátzico confirmed these assumptions. All of the ceramic and lithic artifacts from excavation seem consistent with a low-density Late Postclassic Tarascan occupation. Based on the site typology established by Parsons et al. (1983) for the Valley of Mexico, the variability observed in the area of Itziparátzico may indicate 1) one large, dispersed village; or, 2) an amalgam of several hamlets. The presence of sizable mounds, however, seems to point more toward a dispersed village.

The combination of ceramic pipe fragments, polychrome pottery with resist decoration, and gray-black obsidian prismatic blades (presumably from Ucareo), may indicate presence of Tarascan office holders at this location, as such items have been found associated with Tarascan elite burials (Pollard and Cahue 1999; Pollard et al. 2001). Both ceramic pipes and polychrome vessels with resistant decorations are considered markers of a Tarascan elite identity that was fully defined by the fourteenth century (Pollard and Cahue 1999: 278). It has also been suggested that the Tarascan central dynasty directly controlled the obsidian from the Ucareo mines (Pollard and Cahue 1999: 273). Some of the ceramics represented in the Itziparátzico sample may correspond to types defined by Pollard (2005) at Erongarícuaro, an important Tarascan site in the Pátzcuaro Basin.

According to Colonial documents, the area was in the hands of the Tarascan nobility in prehispanic times. This seems to be supported by both environmental impact assessments and archaeology. Evidence also indicates that primary copper production was being carried out at this location. This is particularly significant since, in the absence of mines in the area, ore had to be carried some 125km. The access to pine-oak uplands where forests would provide high-quality charcoal seems to have been a major factor for locating the smelters here. The analysis of metallurgical remains was the next critical step to more effectively define the nature of the copper industry at Itziparátzico.

Slag analysis

The analysis of slag to obtain technological information on the smelting process represents a critical component of this research project. The use of ores is invariably related to the formation of slags, because these waste products act as collectors for impurities introduced into the smelting process. For this reason, the composition and properties of metallurgical slags are influenced by the following variables: 1) the ores and gangue; 2) the fluxes added; 3) the furnace construction materials (including lining, tuyeres, etc.); 4) the ashes of the charcoal and wood used in the process; 5) the process conditions (heat distribution, air intake, furnace profile and height retention time of slags inside the furnace); 6) cooling conditions; and 7) weathering (dependent on time, climate and deposition). Slag analysis thus has the potential to reveal a range of important information about metallurgical technology (Bachmann 1982).

Slag samples recovered from the excavations at Itziparátzico were analyzed for microstructure and compositional properties using light microscopy, x-ray fluorescence spectrometry (XRF), and scanning electron microscopy with energy-dispersive x-ray spectrometry (SEM/EDS) (raw results in Appendix C). The main goal of this research was to obtain technological information concerning the metallurgical process at Itziparátzico. Several specific questions were addressed:

1. Is the material smelting or casting slag?
2. Is the distinction between macroscopically distinct slags relevant in terms of their microstructure and chemical composition?
3. What are the technological implications of these differences?
4. What type of mineral ore was being processed at the site?
5. Can information be gained about furnace temperature and atmosphere?
6. How efficient was the smelting process?

Two major types of slag were identified macroscopically and labeled as either lumpy or platy, according to their general morphology. *Lumpy* slags are commonly characterized by plano-convex to irregular surfaces, with sizes ranging from gravel-size up to 2kg. Specimens often present smooth surfaces on one or both of their faces; porosity is always apparent, and the fracture is usually uneven (Figure 4.26). *Platy* slags, in contrast, are usually flat, with minor amounts of porosity and signs of ropey flow on one or more surfaces. The thickness ranges from approximately 2 to 9mm. Generally, one of the surfaces exhibits a metallic to glassy finish, while the other shows a more resinous luster. Failure normally produces a conchoidal fracture (Figure 4.27). In both cases the slags consisted of crystalline phases together with a glassy phase. Similar slag morphologies have been observed in Old World copper slags (see e.g. Meyerdirks et al. 2004).

Sample selection, preparation and processing

Nineteen slag fragments were selected for analysis, from a sample of approximately 2.1kg. Sixteen of the selected specimens were recovered from stratigraphic levels excavated at Itziparátzico (Units 1-7; see Appendix C). The other three fragments are part of the surface collection from Jicalán El Viejo, a presumably contemporary site about 90km to the west (see Figure 4.8). The samples from Jicalán were analyzed for comparative purposes, since no references to previous slag analyses from Mesoamerica were available at the time.

All samples were analyzed by optical microscopy and scanning electron microscopy with energy-dispersive analysis (SEM/EDS). Twelve of the sixteen specimens selected for analysis were investigated by bulk chemical analysis using standard XRF procedures. All analyses were performed at the Wolfson Archaeological Science Laboratory, Institute of Archaeology, University College London. Two basic sample preparation methods were used, depending on the analysis to be performed: microscopic analyses (optical microscopy and SEM/EDS) or chemical composition analysis (XRF).

One of the sample preparation methods consisted of cutting sections from the samples using a diamond wheel and then mounting them in epoxy resin. The sections were ground sequentially on 240-, 320-, 600-, 1200-, 2500-, and 4000-grit silicon carbide papers and polished with diamond paste to 1μm. For scanning electron microscopy (SEM) and energy-dispersive spectroscopy (EDS), the specimens were subsequently carbon coated to ensure electrical conductivity. The second method involved the use of the same set of samples for XRF analysis.

About 10g of each specimen was crushed to powder using a tungsten ball mill. The powder was then mixed with wax and pressed into a pellet. A total of fifteen homogenized pellets were prepared and subjected to XRF analysis to determine the bulk chemical composition of the slags, using special calibration methods developed at the Institute specifically for Fe-rich slag.

Microscopic analyses of slag samples

The analysis of the sixteen slag samples began with optical microscopy to determine their mineralogical

Figure 4.26 Lumpy slag fragment from Itziparátzico.   Figure 4.27 Platy slag fragment from Itziparátzico.

composition. The polished samples were mounted on micro-slides for analysis on a reflected light microscope fitted with 5x to 100x objective lenses and a 10x eyepiece, providing magnification of 50x to 1000x. The microscope was fitted with a rotating stage to allow for the identification of individual minerals by use of polarized light. Mineral phases and major elements of the selected specimens were characterized using SEM and EDS on a Jeol JSM-35 scanning electron microscope equipped with a LINK ISIS energy-dispersive spectrometer.

In scanning electron microscopy, the sample is scanned with a highly focused electron beam. The secondary electrons emitted have their origin in small depths of the sample and are used for imaging the surface topography. The resolution is limited by the diameter of the electron beam. The backscattered electrons provide information about the atomic number of the elements in the sample. This image is used for element contrast. The electron beam can be used for excitation of emission lines in the x-ray spectrum. The characteristic x-rays emitted are used for qualitative and quantitative elemental analysis (see Pollard and Heron 1996).

The SEM measurements were performed at an accelerating voltage of 20 kV, a probe current of 1 nA (nanoampere), and a working distance of 10.5mm. Under these analytical conditions, the probe diameter on the sample surface was about 100nm (nanometers). The morphology of the different phases on the sample surface was imaged using backscattered electrons in compositional contrast mode. Quantitative microanalysis was performed using the conventional correction procedure included in LINK ISIS software.

Each sample had up to ten separate points analyzed. Metal and sulfides in the slag were analyzed *quantitatively* for S, Fe, Ni, Cu, Zn, As, Sn and Pb. In the case of oxide phases, the following elements were measured *quantitatively* and combined by stoichiometry with oxygen: Na, Mg, Al, Si, P, S, K, Ca, Ti, Mn, Fe, Ni, Cu, Zn and Pb. In both cases, the final results were normalized to 100% to compensate for porosity and instrumental drift. Individual reports of the SEM analysis for each sample are presented in Appendix C.

Summary results of the microscopic analyses

The dominant phases in both slag types were fayalite, quartz, spinel, glass and various sulfides, together with small prills of virtually pure copper. Figures 4.28a and b are low magnification light micrographs, which were recorded from the same 'lumpy slag' specimen. A duplex copper/copper sulfide prill (C) is seen in association with a gas bubble (D) in Figure 4.28a (and see Figure 4.30b). Plate-like crystals of fayalite (Fe_2SiO_4) are seen in regions of the slag (labeled A on Figures 4.28a, b) that had been fully molten.

The lumpy slags contain a high concentration of only partly reacted quartz grains, and one such region is labeled 'B' on Figure 4.28a. In contrast the 'platy slags', contained little or no un-reacted 'host-rock'. We were also able to unambiguously identify the mineral ore as chalcopyrite ($CuFeS_2$) because pockets of the mineral were still present in the slag (e.g., at E on Figure 4.28b). Figure 4.29a is an SEM, backscattered electron image of the same region of chalcopyrite as that shown in Figure 4.28b, whilse Figure 4.29b is an EDS spectrum from the central portion of Figure 4.29a: the iron, copper and

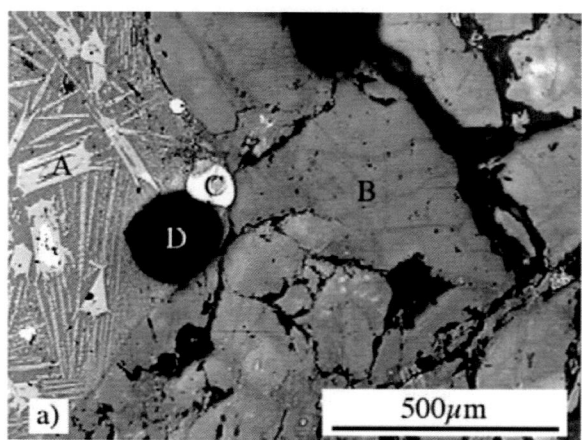
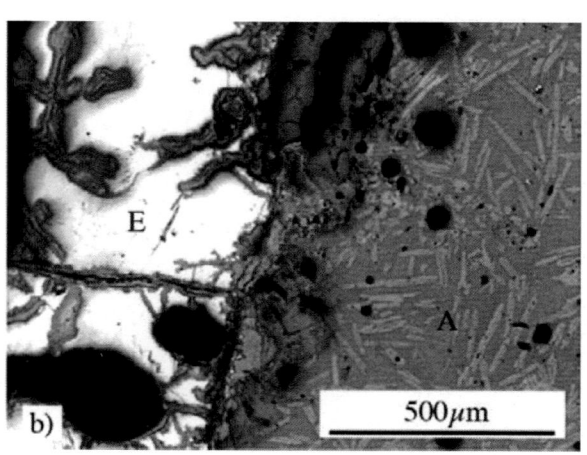

Figure 4.28 Low magnification images of a typical ‹lumpy› slag sample.
a) Region of slag that contains a fayalitic solidification structure (at A), together with a polycrystalline aggregate of un-reacted quartz grains (at B). A duplex copper/copper sulfide prill (C) is seen in association with a gas bubble (D). b) Regions of the slag that contain fayalitic slag (A), together with an un-reacted aggregate of chalcopyrite (E).

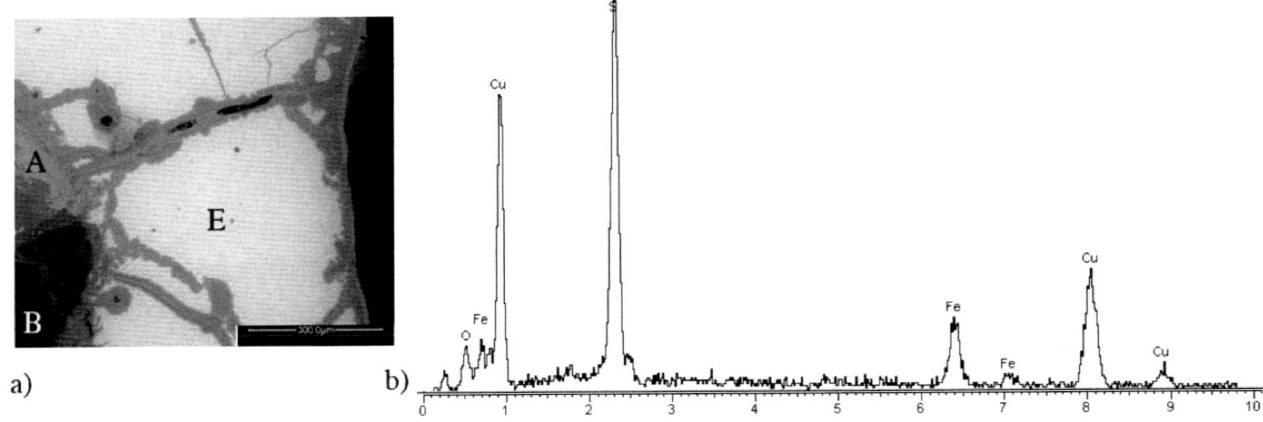

Figure 4.29 Identification of the copper ore: chalcopyrite.
a) Backscattered SEM image of the same region (E), as that shown in Figure 2b. Fayalitic slag is present at A, and un-reacted quartz is found at B. b) Energy dispersive x-ray spectrum from the chalcopyrite in Figure 3a. The copper, iron and sulfur peaks are labeled.

sulfur peaks are labeled (see Maldonado et al. 2005). Figures 4.30a and b represent SEM, secondary electron, images of the fayalitic slag microstructures, which were common to both slag types. The crystals of fayalite are plate-like, and their distribution is strongly suggestive of eutectic solidification; that is, it may have involved a mixture of two or more phases solidifying at a particular temperature and particular composition of each phase (see Cottrell 1995: 228).

EDS analyses of the matrix regions reveals (in addition to oxygen), the presence of iron and silicon, together with lesser amounts of aluminum, calcium and potassium, i.e., the matrix is probably a glassy phase that has partially devitrified. Figure 4.30b also shows a common occurrence in both slag types: duplex copper/copper sulfide prills. The core of the prills is virtually pure copper, while the annulus is invariably a copper sulfide. The regions of the slag samples that had been liquefied are all fayalitic, with iron oxide and silica jointly averaging at least 80 weight percent. Alumina accounts for c. 5 to 10 weight percent, and lime, magnesia and potash for one to two percent each (Maldonado et al. 2005).

Chemical composition analysis of the slag samples

Bulk chemical analyses were performed by X-ray flourescence spectrometry (XRF) for twelve out of the sixteen slag samples. This analysis used a Spectro Xlab 2000 and evaluated the measured values against certified reference materials for quantitative analysis. For this purpose, a special slag calibration curve was used (see Veldhuijzen 2003). X-ray fluorescence is the excitation of emission lines in the X-ray spectrum by X-rays. Electrons on a level close to the atomic

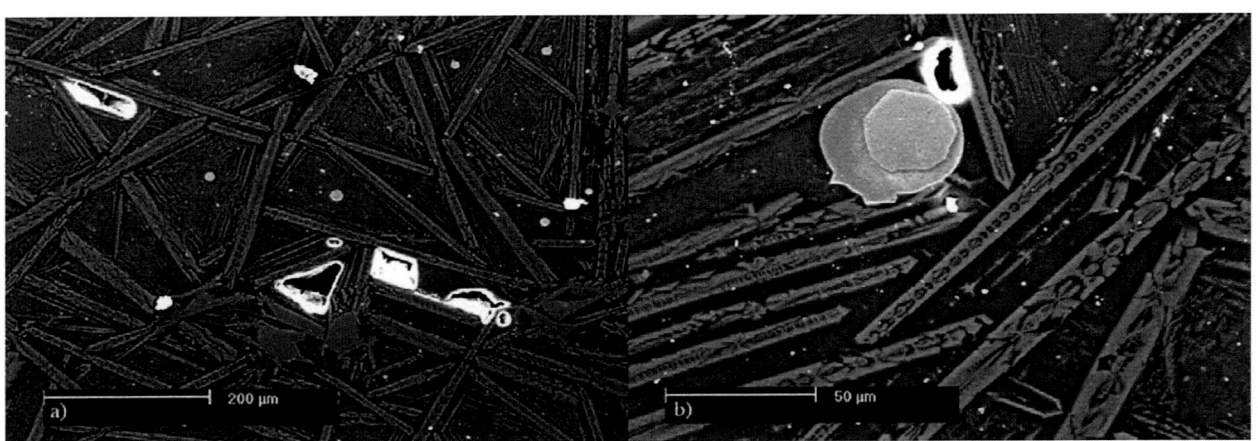

Figure 4.30 Scanning electron microscope (SEM) images of a ‹typical› region of a lumpy slag.
a) Low magnification image. The network of fayalitic plates (light gray) is typical of eutectic solidification. Duplex copper/copper sulfide prills are observed towards the centers of the unresolved eutectic mixture. Shrinkage cavities are also observed. b) Higher magnification images of fayalite plates, and a duplex copper/copper sulfide prill. EDS analyses showed that little iron was associated with either the copper or the copper sulfide. The 'unresolved' matrix regions contained FeO, SiO2, Al2O3, CaO and K2O.

nucleus are excited. As they return to their initial level, the energy difference is emitted as characteristic X-radiation. This radiation is analyzed by the detector system for intensity and spectral distribution (Pollard and Heron 1996).

Smelting slags are composed primarily of silicates, fayalite (iron olivine, Fe_2SiO_4) being the most common component in early copper smelting slags. Fayalite formation relies on the presence of sufficient iron and silica in the slag (Bachmann 1982; Merkel and Rothenberg 1999). The compound fayalite has a melting point of 1170°C and can absorb small amounts of MnO, MgO, Al_2O_3 and FeO, which may raise or lower its melting point by up to 50°C (Tylecote 1962: 187). In the slag assemblage from Itziparátzico, fayalite occurs as intergrowths with magnetite (Fe_3O_4) and occasional irregular shaped prills (see Figure 4.28b).

Although relatively consistent in terms of their phases, the slags from Itziparátzico show a certain degree of heterogeneity, presenting various partially reacted phases, minerals and metal. Two major slag 'types' with different chemistries were noted in the analysis, which correspond closely to the macroscopically identified 'lumpy' and 'platy' slags (Figures 4.26, 4.27). Platy slags tend to have c. 45 weight percent FeO and 30 to 40 weight percent SiO_2, while lumpy specimens have only c. 30 weight percent FeO and 50 to 60 weight percent SiO_2. Two samples, 1-3b and 3-1b, fall halfway between the two groups (see Figure 4.32).

Summary of the results of the chemical composition analysis

The much higher silica level in the 'lumpy' group reflects the presence of un-reacted quartz particles (e.g., see Figure 4.28a), which 'dilute' the slag composition as compared to the platy slags. Most minor oxides show the same (though variable) trend as iron oxide, with only about 2/3 their platy slag values being found in the lumpy slags. Exceptions are alumina, lime and potash. All three are more variable in their concentrations in the lumpy than the platy slags, and do not appear to be diluted by the 'un-reacted' silica. Moreover, alumina, potash and, to a lesser extent, lime co-vary. It may well be that the quartz-rich host-material contained feldspar or similar minerals: a suggestion that is reinforced by the presence of aluminum, calcium and potassium in the matrix regions of the fayalitic slag. Possible rock types providing such an assemblage include arkose sandstone and igneous intrusions such as those often associated with porphyry copper ore deposits; however, more geological research is necessary to document the source of the copper ore and its host rock (Maldonado et al. 2005). The chemical analyses also indicate an elevated level of sulfur in the lumpy slag (expressed in the table as SO_3, but present in the slag as sulfide, S^{2-}). This is consistent with the observation of chalcopyrite ($CuFeS_2$) in the microstructural analyses of the lumpy slag (e.g., see Figures 4.28b, 4.29a and 4.29b).

Bulk analyses derived from XRF analyses can also be plotted on a phase diagram, as illustrated in Figure 4.31. The evaluation and interpretation of phase diagrams is a sophisticated and rather complex field of physical chemistry that is further explained in Appendix C. Nevertheless, a few explanations for basic understanding can be provided. Phase diagrams are commonly used to study the relationship between chemical composition and compound formation during and after the solidification of slags (see Bachmann 1982: 10-13).

CONTEXT	SLAG	SiO2	Al2O3	FeO	TiO2	MnO	CaO	MgO	Na2O	K2O	P2O5	SO3	CuO	PbO	ZnO	TOTAL
		%	%	%	%	%	%	%	%	%	%	%	%	ppm	ppm	%
1-1c	p	34.1	7.20	49.7	0.23	0.06	0.63	1.88	0.36	0.92	0.05	0.13	2.41	0	500	97.4
1-2b	p	32.4	10.86	46.9	0.35	0.13	1.46	2.06	0.56	0.82	0.11	0.27	1.29	225	1050	97.2
1-4a	p	35.0	9.65	43.7	0.29	0.09	1.72	2.30	0.61	1.11	0.07	0.26	1.48	0	650	96.2
1-4c	p	35.2	6.40	50.3	0.21	0.09	2.44	1.62	0.58	0.82	0.09	0.59	0.73	400	700	99.0
2-1b	p	40.0	3.44	46.2	0.04	0.13	4.29	0.61	0.56	0.50	0.01	0.24	1.44	1500	1100	97.7
1-2a	l	56.4	7.78	30.5	0.27	0.07	0.89	1.29	0.41	1.47	0.05	0.48	1.57	0	150	101.0
1-3a	l	55.9	7.59	30.3	0.23	0.06	1.71	1.30	0.44	1.34	0.05	0.42	1.23	0	150	100.5
1-3b	l	41.2	7.56	42.8	0.25	0.09	0.93	0.96	0.56	0.96	0.10	0.46	1.11	380	1350	97.0
2-1a	l	54.4	11.06	27.5	0.23	0.08	2.50	1.44	0.35	1.95	0.04	0.71	0.77	0	250	101.1
3-1a	l	57.0	5.04	32.7	0.15	0.06	0.59	0.52	0.21	0.84	0.03	0.35	1.87	400	600	99.4
3-1b	l	41.6	7.12	44.0	0.24	0.07	2.23	1.46	0.57	1.06	0.07	0.71	3.28	150	2800	102.5
3-1c	l	61.7	3.96	32.6	0.12	0.04	0.48	0.03	0.08	0.87	0.06	0.54	0.93	300	2550	101.4

Figure 4.31 XRF analysis of copper slag from Itziparátzico.
In the slag column, 'p' stands for platy and 'l' for lumpy, referring to the two major types of slag identified in the sample.
Analyses were done on powder pellets using a calibration method set up by Veldhuijzen (2003).

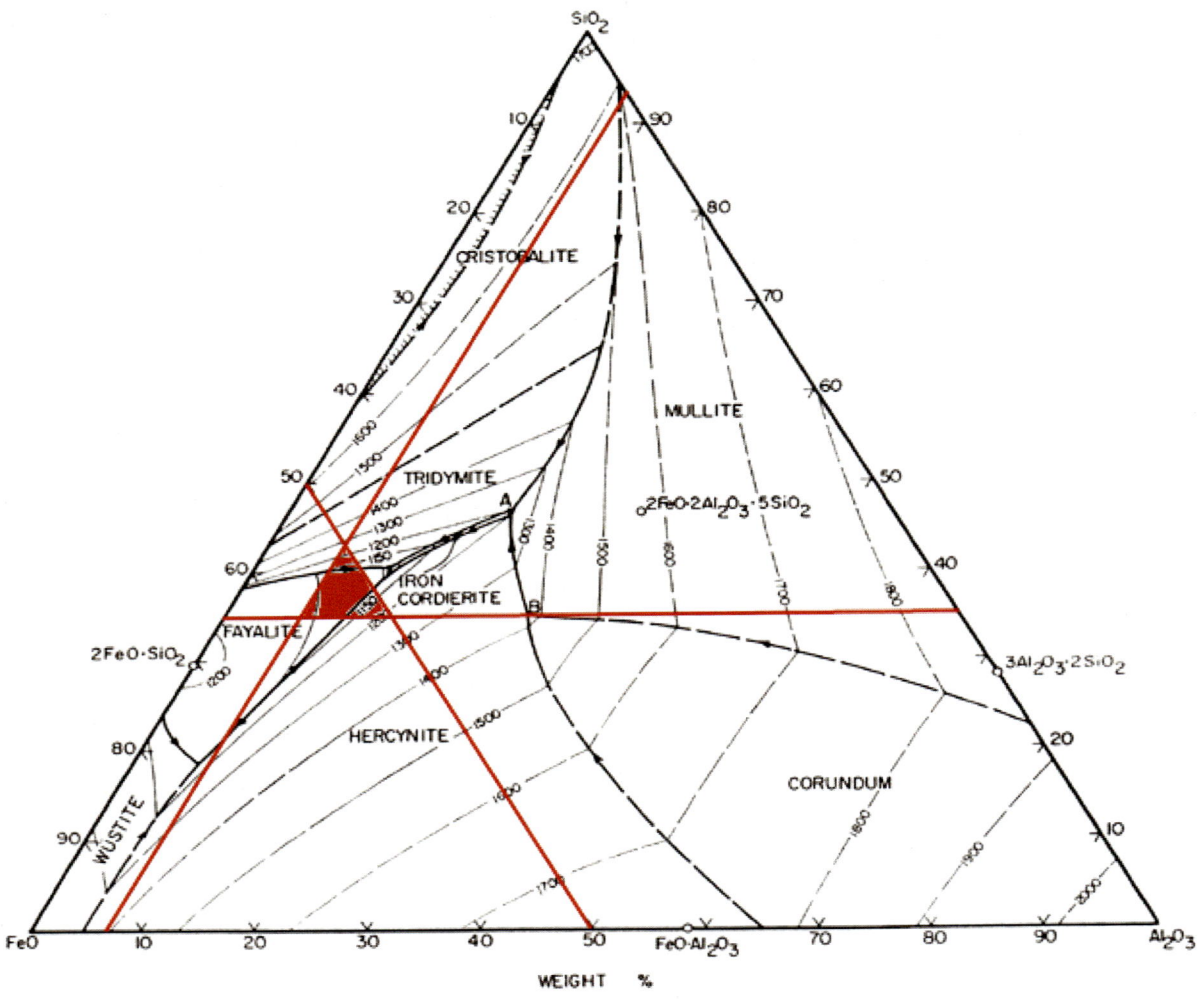

Figure 4.32 Triangular diagram showing phase relationships in the system iron oxide- Al2O-3 SiO2 under reducing conditions (after Muan 1957: Fig. 10). The plotted points correspond to the values obtained for sample 1-4c (see Table 4.2) and fall roughly within the boundaries of the fayalite phase field.

As observed in Figure 4.32, the major oxide components of the slags from Iztiparátzico are FeO, SiO_2 and Al_2O_3. In order to illustrate the phase relationships in a simple way, the compositions can be projected into a triangular plane, and the diagram thus obtained has the appearance of a ternary system. Such a diagram then necessarily gives a somewhat distorted picture of the phase relationships, but it is useful and valuable as long as the true relationships (weight percent values) clear. In the present case, a fayalitic phase has been identified through microscopic examination.

The ternary phase diagram for the system iron oxide- Al_2O_3- SiO_2 shown in Figure 4.31 indicates that fayalite begins to form at about 1100°C from a reaction involving silica and iron oxide under reducing conditions. A number of variables in the system should, however, be be taken into account. First, the data include free silica from residual quartz. While this compound would fall in an area of unrealistically high melting temperature in the diagram, the presence of unreacted free silica disproves this notion. A second factor to bear in mind is that not all the iron in the slag is present as FeO; some of it may be present as Fe_3O_4 (magnetite), which will have subtle effects on the system as well.

Interpretation of the Scientific Analyses of the Slag Samples

The identification of un-reacted chalcopyrite, together with prills of copper sulfide and copper, unambiguously characterize the slag from Itziparátzico as copper smelting slag. The bulk composition and the ubiquitous presence of fayalite match other known examples of fayalitic copper slag. The mineral ore being processed has been shown to be a sulfidic ore, chalcopyrite ($CuFeS_2$), in a silica-rich matrix. The dominance of fayalite and the presence of copper prills indicate: 1) a high smelting temperature, of around 1100°C; and 2) the occurrence of smelting under strongly reducing conditions. A reducing atmosphere under such high temperature is normally inconsistent with mouth-blown smelting operations. While the maximum attainable furnace temperature with blowpipes was sufficient to smelt copper, the combination of temperature and reducing atmosphere necessary for fayalite formation are unlikely to occur using this system.

The physico-chemical conditions of preindustrial smelting processes were controlled by the relative amount of charcoal used and air blown into the charcoal bed. The composition of the mineralogical phases in slag therefore reflects the ratio CO/CO_2 as a measure of redox conditions (Hauptmann et al. 2003). The chemical composition of human breath is such that when it is used to burn charcoal in a well-insulated and preheated furnace, it reaches a maximum temperature of about 1200°C, although this can be highly variable since the composition of breath varies for physiological reasons.

In general, however, the oxygen content of human breath decreases and its CO_2 content increases as blowing effort increases. The predominance of CO_2 over CO promotes an oxidizing (rather than a reducing) atmosphere. In addition, lower oxygen content and the presence of CO_2 and water vapor change the chemistry of the combustion reaction, decreasing the heat generated. If furnace temperature is decreased appreciably (by a lower blowing effort, for example) the slag can become viscous and retain smelted copper in the form of disseminated small prills or pellets (Rehder 1994). The slags had to be broken up to recover the prills of copper. Such evidence is commonly observed in early copper smelting in South America (e.g. Shimada et al. 1982) and in the Near East (e.g. Rothenberg and Merkel 1995), but not at Itziparátzico.

Despite the observed inclusions of partially decomposed materials in the slag from Itziparátzico, the predominance of fayalite indicates reducing conditions, while the overall microstructure of the slag suggests that the primary materials fully reacted to form a relatively homogeneous melt. Relatively low copper losses are attested from the bulk composition measurements. The presence of un-reacted host materials in the lumpy slag is particularly informative as it reflects part of the smelting process.

The results of the microscopic analysis show that there is little un-reacted material in the platy slag. Thus, platy slags appear to represent a more advanced stage and lumpy slags seem to belong to an earlier stage in the extractive process. The two slag types may thus represent sequential waste products of the same continuous smelting process, possibly relating to consecutive tapping events (Maldonado et al. 2005). As pointed out in Chapter 3, the smelting ore in a reducing environment converts the solid gangue into a liquid (slag) so it can be tapped off, while at the same time it allows the metal to form and separate.

The presence of tap slag is typically associated with a smelting process carried out in a furnace rather than a crucible (Craddock 2001: 157). Tap slag is formed when, upon completion of smelting, molten slag is allowed to run out of the furnace, after which it cooled rapidly (see Bachmann 1982: 4; Rothenberg 1985:126). Tapping is characterized by a number of features observed in the slag samples. They tend to show clear flow textures indicating the top surface. The ropey flow shape in the more viscous slags (i.e lumpy slags) may indicate that they formed close to the tapping hole, whereas slags of lower viscosity and a plately form (platy slags) may have solidified in the collecting area. In some specimens the top surface is highly oxidized, while the bottom presents impressions of sand grains from the soil over which it flowed and a much less oxidized surface. Rapid cooling is revealed by the high glass content in many of the samples.

Discussion

The socio-technological summary in this chapter provids the necessary background information on the Tarascan state, for the consideration of the interrelations of technological, economic and political organization discussed in the following chapters. The character of the regions's environment (its topography, climate and natural resources) has a close relationship to the history of both settlement and technology. The influence on the general patterns of settlement in the Santa Clara-Opopeo region is clearly beyond the scope of this study and has yet to be adequately explored.

The physiographic summary presented here is simply an outline of some of the major environmental factors that bear on the location of smelting areas and thus the organization of Tarascan metallurgy. Environmental data, however, support ethnohistorical accounts that the lands in most of the Basin of Zirahuén were in the hands of the Tarascan nobility, which places the smelting processes carried out at Itziparátzico in an elite context. Finally, generalizations have been drawn from the description of the fieldwork methodology, categories of data collected, and features of the copper production deduced from the patterns of distribution and general characteristics of such data, to determine the nature, scale and organization of the copper industry identified in the area.

The evidence found at Itziparátzico has allowed us to conclusively determine that the production activities carried out there involved only primary smelting, since no other metallurgical activity was recorded during the investigations. Copper ore, positively identified as chalcopyrite, was brought to Itziparátzico for smelting. Copper ingots were then probably transported elsewhere for their final processing into objects. The evidence of tap slag at Itziparátzico has extremely important implications, as it suggests that the smelting process took place in a furnace rather than in a crucible. These data raise important questions about the chronological context and development of the primary production of copper in the area. The technology employed in the process involved an efficient, highly reducing smelting environment, which would be difficult to achieve using lung-powered blowpipes and crucibles (see Rehder 1994). If the smelting activities at Itziparátzico involved the use of furnaces powered by hand-worked bellows, we may well be dealing with a relatively early post-Contact production area that largely overlapped with prehispanic traditions, as indicated by the evidence of Tarascan pottery and lithics.

Alternatively, if the operations at the site are Late Postclassic in date, this would indicate that Tarascan metalworkers had developed a copper smelting method which involved some form of natural draft. The exposed position to prevailing strong winds in this area (especially during the dry season), have led me to consider that a naturally wind-aided processing may have been taking place at the site. The use of wind power for smelting has been suggested for a number of sites in the Old World and has also been identified in pre-Columbian South America. This technological possibility is further explored in Chapter 6.

The reference to indigenous smelting specialists from the town of Opopeo in an early Colonial source (i.e. (*Archivo General de la Nación* [AGN], *Reales Cédulas Duplicadas*, v. 16, exp 346, f. 108, in Barrett 1987: 23) seems to support the existence of a native copper industry in the area. In either case, it is a fact that copper ore (chalcopyrite) was brought to Itziparátzico for smelting, and copper ingots were probably transported elsewhere for their final processing (Maldonado et al. 2005). The mineralogy of chalcopyrite is consistent with the sulfidic ore bodies that comprise the mining district of La Huacana (Carranza A. et al. 1995).

Chapter 5

Models of technological organization

Numerous questions arise from the notion that metal items functioned as wealth finance in the economy of the Tarascan state of Late Postclassic Michoacán. Foremost among these is whether and how wealth was produced and controlled by the central power. Metallurgy, however, does not represent a single technological process. The transformation from ore to finished product involves many individual stages and numerous choices have to be made throughout the production sequence. The metallurgical *chaîne opératoire* (Leroi-Gourhan 1963; see Chapter 2) for Tarascan copper production requires a specific body of expertise and skills that includes: the procurement of minerals from which to extract metal (extractive metallurgy); the identification and use of alloys; the working of metal through some mechanical method (hot or cold hammering) or by fusion and molding (foundry); the application of polishing and other finishing techniques; and the knowledge of the required forms and symbols.

Several important issues regarding Tarascan metallurgy arise from the above facts. How many different categories of producers participated in the manufacture of metal items in the Tarascan Empire? What was the degree of specialization of these producers in their particular craft? What was the level of involvement of the central power of the state and the elites in each one of the stages of production of this craft? Although current data for both mining or extractive metallurgy in Mesoamerica are sparse and unclear, important aspects of the operational sequence for Tarascan copper production, including ore deposit and mining, smelting, and final processing, can be inferred from a combination of ethnohistorical and archaeological data. The present chapter uses a political ecological approach (see Chapter 2) to explore some dynamic levels of interaction between the production of copper and copper-based goods and the centralized power of the Tarascan Empire of Late Postclassic Mesoamerica. Archaeological evidence and documentary sources provide a picture of organizational patterns in the prehispanic Tarascan copper industry.

This chapter is organized in three major parts. First, a brief discussion of the theory and methods used by archaeologists in the study of craft production is presented. Second, two possible political ecological models of technological organization are suggested as relevant to the interpretations of the data on copper production at Itziparátzico. As idealizations of reality against which the data may be compared, the models outlined here are primarily heuristic, intended to provide a way of interpreting the data through generalization and the identification of the distinctive features of prehispanic metallurgy in Mesoamerica.

The rich ethnographic record of African metallurgy is drawn upon to illustrate these models in the second part. The models presented characterize only two of many possibilities within an almost continuous range of variation of possible organizational forms. They are simply two points on this spectrum, chosen for both their generality and their applicability to the conditions of Tarascan metallurgy. Third and finally, based on these two formulations a tentative model is proposed for the organization of the copper industry in the Tarascan state of prehispanic western Mexico.

The study of craft production

As discussed in Chapter 2, the study of specialized craft production (its emergence, organization and associated technologies) has a long history in archaeological research (see Wailes 1996; Patterson 2005). Because much craft production leaves distinctive material traces, archaeologists are able to address technologies, locations and scales of production, and to examine their roles in the broader social, economic and political contexts in which they occur. As mentioned previously, 'political economy' is defined broadly as the relations between political structures and systems and the economic realms of production, consumption, and exchange (Stein 2001: 359; see also Chapter 2). The term craft production entails 'the investment of labor by (more or less) skilled practitioners who labor to transform potential into finished products that were in turn consumed by non-producers' (Sinopoli 2003: 1, 16).

Archaeologists and social theorists have long been concerned with the relationship between craft production and political and institutional structures (Wailes 1996; Patterson 2005). Economic differentiation and political complexity are often seen as being inextricably intertwined and are widely perceived as the defining characteristics of early states and empires. Craft products can perform a number of functions in political economies: as exchange goods, as sources of wealth, as tools that produce the material infrastructure of complex societies, and as prestige goods or symbols of power and status.

The relations between craft producers and institutions tend to be both multiple and multidimensional, with varying degrees of autonomy and interdependence, and much production can occur outside the control of the state (Sinopoli 2003). Two main issues relevant to the study of the political economy of complex societies are often addressed in archaeology: the relations of artisans to institutions, and the roles of craft goods in political economies.

Some notes on attached vs. independent craft production

The distinction between attached and independent specialization suggested by V. Gordon Childe (1951a, 1951b; see Appendix D) has been taken up by many scholars seeking to explore the political dimensions of craft production activities (e.g. see contributions in Wailes 1996; Brumfiel and Earle 1987; Costin and Wright 1998; also Peregrine 1991). In an influential work, Brumfiel and Earle (1987) attempted to further categorize craft production by highlighting factors that link political processes with productive organization. They emphasized the difference between the production of wealth goods and staple products (see also D'Altroy and Earle 1985), and the divergent roles that these played in political economies. In their discussion, Brumfiel and Earle list multiple parameters or dimensions of variation that are embedded in the concept of specialized production, including: the affiliation of specialists (i.e. independent or attached), the nature of the product, the intensity of specialization (i.e. part-time or full-time), the scale of the production unit, and the volume of output (1987:5).

Costin (1991, 2001) expanded on Brumfiel and Earle's perspective in an important study in which four parameters for documenting craft production are outlined: concentration, scale, intensity and context. Costin's focus, however, was on the social, spatial and political settings and structures of production, rather than the goods produced and their uses. She viewed each of the parameters as a continuum that could vary independently. Thus, rather than a limited set of types, craft production could be structured in a multitude of ways, within a single culture and cross-culturally. Costin's parametric approach to the study of craft production provided a valuable framework for disarticulating the different activities and spatial and social contexts of craft production activities. Each of the parameters, however, poses challenges to archaeological research, and not all aspects are easily accessible from archaeological evidence.

The parameter 'concentration' describes the spatial distribution of specialists across the physical landscape in relation to other producers and potential consumers (e.g., rural versus urban, degree of nucleation). 'Scale' refers to both the size of production units and the rules or principles for their recruitment. 'Intensity' is defined as 'the amount of time producers spend on their craft' (Costin 1991:16), with full-time specialization at one extreme of the continuum and part-time specialization at the other. In the latter, craft production represents only one of a set of productive activities engaged in by specialists, while for full-time specialists, producing craft goods is their primary or exclusive economic task. The fourth parameter, which Costin termed 'context' refers to the relationships of productive activities and producers to sociopolitical institutions and/or patrons.

At one end of this continuum is what Costin, following Brumfiel and Earle (1987), referred to as 'attached specialization', in which institutions or elites directly regulate some aspects of craft production. Such regulation may include control over raw materials or production tools, control over access to and/or the quality of finished products, or control over producers themselves. These producers may, incidentally, labor as part of elite households or in institutionally sponsored and supervised workshops. At the other end of the continuum is 'independent specialization'. Here, production is not directly controlled by institutional authorities or elites, but is regulated by autonomous artisans or groups of artisans who work largely independently of political institutions and whose production activities are influenced by consumer demands and competing economic priorities. Independent producers may be involved in political economies (e.g. they may be required to pay tribute related to their production activities), but they have greater flexibility in the scheduling and organization of their productive activities and the disposition of their products than do attached specialists.

Gradients of variation in Costin's scheme

Attached specialists are viewed as being dependent upon, and tethered to institutions. Independent specialists on the other hand, exercise greater choice over their activities and labor. While dualism appears inherent in the scheme, a considerable variability can lie within each of these relations. Thus, attached specialization may include: 1) intensified household production in which kin-based units produce in excess of their needs to satisfy institutional demands; 2) dispersed corvée production, in which a portion of producers' production activities in local settings is reallocated to the state; 3) individual retainer production, where individual skilled artisans are recruited to produce for elite patrons or state institutions; 4) nucleated corvée, in which part-time labor is recruited by the state to work in administered settings or facilities; and 5) retainer workshops, large-scale, spatially segregated workshops with full-time artisans (Costin 1996: 211). Costin's concept of context has generated more discussion than any other topic she discussed, and the characterization of attached and independent specialization, along with its archaeological identification, have proven to be very complex (Stein 1996).

While Costin envisioned a wide variety of relations between sociopolitical institutions and producers, others have been more rigid in their application or conception of the attached vs. independent distinction. As a result, the concept has been subject to considerable criticism. In particular, the distinction between these two categories has been criticized for not adequately addressing the wide variety of relations that can exist between producers and patrons (e.g., Inomata 2001; Janusek 1999). Such relations include, as Ames (1995), Helms (1993), Inomata (2001), Reents-Budet (1998) and others have pointed out, the fact that members of elite households and status groups may themselves be craft producers of high value and restricted prestige goods. The recurrence of this critique in the literature suggests that there is something problematic in how archaeologists have conceived of, or applied Costin's original formulation. The problem seems to lie in a tendency to perceive attached and independent specialization as 'states of being' or fixed and exclusive relations, rather than as particular relations of production with specific, and often limited, duration (Sinopoli 2003: 32-33; Stein 1996: 25-26).

In recent years, archaeologists have found it necessary to refine this model in order to apply it to specific research cases. Janusek (1999), for example, has proposed the term 'embedded specialization' to define the organization of production at the level of corporate kin groups in Tiwanaku. Embedded specialization entails relations of production that are more centralized than independent specialization, but less constrained than attached specialization. In this context of production we may find a relatively high number of individuals who do not fully participate in subsistence activities because they can depend upon their kin ties. Access to raw materials also tends to be higher than that found among societies with a more 'classic' system of attached specialization. The *calpulli* organization in the Aztec capital of Tenochtitlan may represent an example of this form of specialization. In other instances, however, elites may have had control over the distribution networks or the means of transport, so independent specialists relied on them for the distribution of the goods produced (see Arnold and Munns 1994).

The above suggests that sometimes, autonomous production which reaches a level higher than that of local consumption needs, has to depend on a ruling system for the distribution of the goods produced outside the community. The state-controlled merchant organization of the Aztec *Pochteca* (Hassig 1985) may represent an example of this form of semi-dependent specialization. In some cases, individual households may also display high levels of specialized production. Mesoamerica provides numerous examples of cases where individual households could have access to raw materials and skills to produce goods on a relatively large scale. For example, Feinman (1999) has argued that in early state periods in Oaxaca, many craft products were produced in household contexts. The evidence for such situations challenges our notion of the relationships between scale and intensity in craft manufacture (Feinman 1999; Feinman and Nicholas 2000). These examples should lead us to question the idea that large-scale production appears only among complex societies participating in trade networks, which foster independent or attached production away from the household context.

It is important to recognize as well that even in formal workshop contexts, the kinds of relations that may have existed between producers and institutions can vary widely, and can change over time. For example, artisans in the Chinese imperial porcelain kilns at Jingdezhen produced ceramics for the imperial court for much of the year, and during those periods potters were supervised by and 'attached to' imperial officials. When the official demands were filled, the Jingdezhen potters produced similar wares for market distribution, using the same kilns, technologies and personnel (Hobson 1962; Hayashida 1995:17, cited in Sinopoli 2003: 34). Archaeologically, the material signatures of kilns and production debris would be identical for the periods of independent and attached production, as would many of the ceramic products manufactured (though they would differ in the presence of an imperial mark on the vessel base). The political and economic contexts of ceramic production were, however, radically different (Sinopoli 2003).

Other examples might include textile production by Aztec women, who devoted a portion of their efforts to meeting obligatory tribute demands, while also producing goods to meet the needs of their household and to distribute through market or other mechanisms. Whoever they produced for, production occurred in household settings using a consistent technology. Only the intended consumer varied (Brumfiel 1991). Aztec economy in this regard, seems to also exemplify what Stein (1996) has called a dual economy, in which certain specialists independently produced utilitarian items for the market, but were simultaneously attached to the elites and produced prestige and utilitarian goods for them (see Sahagún 1969-82 Bk.9, Ch. 20). This example challenges our notion of a dichotomized independent as opposed to attached specialization, since we may find attached specialists who, contrary to our expectations, were producing utilitarian goods along with (or instead of) wealth goods.

In sum, it is evident that despite the variability encompassed in the concept of 'attached specialization' and its opposite, 'independent specialization', these remain valuable parameters to consider when examining the relations that craft producers maintain with institutions. It is also clear, however, that categorizing evidence of production as corresponding to either attached or independent specialization is not sufficient to describe the precise nature of the relation that existed between specific producers and patrons (Sinopoli 2003).

Scale of production units

As noted by Costin (1991), craft production can occur at a variety of scales and in a range of productive contexts. Costin's 'scale' parameter describes the size and organization of production facilities. Other studies published over the last few decades have also included important typologies. Van der Leew (1977, 1984) presented a five-part classification of craft production units. Although focused on ceramics, his framework was intended to be adapted to other spheres of craft production. Criteria for distinguishing among van der Leew's five types include size of the unit of production, the intended consumers, and the technology of production.

The most basic level in this hierarchical classification is 'household production', in which individuals produce for consumption within their own household, and in which all households in a community produce the same range of goods. While non-specialized in the sense in which the term is most often used in archaeology, household production is minimally characterized by divisions of labor according to gender and age, such that not every household member engages in identical productive activities.

Formal specialization appears in van der Leew's category of 'household industry.' Production still occurs in the household context, but the unit of consumption has expanded beyond the producer's household to include other households in the community. Specialists typically have added craft production activities to their normal range of household tasks, although craft production does not provide their primary economic support.

The third level of production is the 'workshop industry.' Artisans in these contexts derive the bulk of their income from craft production, and produce at a larger scale and on a mostly year-round basis. Workshops may be spatially separated from households or may be located in household contexts. Often, nuclear or extended families comprise the basic unit of production, although some recruitment of non-household laborers may also occur. Some task specialization may occur among workshop members. The scale of production may be quite variable, as may the amount of effort devoted to craft production versus other economic activities by individual workshop members.

Fourth is what van der Leew has termed the 'village industry.' This is characterized by multiple producers and workshops located within a settlement, who produce to serve the needs of a larger region. Production may be on a relatively large scale, with corresponding development of task specialization, so that the various steps in the production processes are themselves specialized and linked to one another (van der Leew 1984: 756). Village producers may be associated with merchants or traders who distribute craft products beyond the village bounds, or may be directly involved in distribution themselves.

Van der Leew's village industry has some similarities to the category 'village specialization' or 'community specialization' that has also appeared recently in the literature (e.g. Hegmon et al. 1995; Welsch and Terrell 1998). Hegmon et al. (1995:33) define community specialization as a context 'in which individual specialists, aggregated in a limited number of communities, produce pottery [or other goods] for regional distribution.' In this case, while people in certain villages may 'specialize' in the production of specific wares, the scale of production may still be relatively low, with production occurring in domestic contexts as household specialization or household industry. There are thus some significant differences from van der Leew's village industry category, which encompasses both the location of production and its scale.

The 'factory industry' is the fifth and largest scale form of production in van der Leew's typology of productive types. Here, production occurs in workshops or factories that are physically isolated from household contexts, and involves large numbers of individuals. Often, these individuals are not related through kin ties, and may be recruited by a variety of means (wage labor, corveé labor, enslavement, etc.). In factory industries task specialization is high, as is investment in production facilities and raw materials. Emphasis on product standardization and quality control may also be high, and such facilities typically include managerial personnel responsible for coordinating, scheduling and evaluating production. These managers can include state or other administrative personnel (e.g. temple elites, military officials), merchants, or private entrepreneurs.

Similar typologies of scale of production have been developed by Peacock (1982) and Costin (1991). Like van der Leew's classification, their approaches are hierarchical and increase in scale from individual

Van der Leew (1977, 1984)	Peacock (1982)	Costin (1991)
Household Production	Household Production	Individual Specialization
Household Industry	Household Industry	Dispersed Workshop
Workshop Industry	Nucleated Workshops	Community Specialization
Village Industry	Manufactory Production	Nucleated Workshops
Factory Industry	Factory Production	Dispersed Corvée
	Estate Production	Individual Retainers
	Military and other Official Production	Nucleated Corvée
		Retainer Workshop

Figure 5.1 Typologies of the organization of craft production (adapted from Sinopoli 2003: Table 2.1).

specialization at the household level to large-scale production in factory contexts, with associated increases in technology, time investment and output. Costin and Peacock's classifications differ from van der Leew's in their incorporation of information on managerial control into the analysis of the scale of the unit of production in their definitions of types (Figure 5.1). While these additions are useful in suggesting who controls or directs craft production in particular contexts, it seems more convenient for the purposes of this work to avoid mingling these different dimensions and to adhere instead to van der Leew's more restricted typology. These various typologies of craft production organization (see also Rice 1991; Sinopoli 1988) are useful in providing ways to think about questions of scale and the social units involved in specialist production. They are also valuable as a frame for discussion, like numerous other simplifying typologies used in archaeology. Nevertheless, like other typologies, they can lead scholars to ignore or reject other possible ways of organizing human labor (e.g. Feinman and Neitzel 1984; Rice 1991; Yoffee 1993). No doubt, numerous other approaches or definitions of units are possible.

What was the scale of production at Itziparátzico?

Based on the data currently available, it is difficult to assess the scale of metallurgical production at Itziparátzico. No smelting facilities were located during the survey or test-pitting, although the highest slag concentrations are located away from what appears to be the residential terraces. The elemental raw materials (copper ores) were not locally available and had to be imported. The technology employed in the smelting process involved an efficient, highly reducing smelting environment, which suggests a complex organization of the craft. This would support the idea that production was in the hands of highly skilled smelters who were part-time to full-time specialists. Pending further archaeological investigations, three of van der Leew's levels of production may apply to Itziparátzico: the household industry, workshop industry, and village industry. All three of these organizational types have been suggested for the production of different crafts in Mesoamerica (examples below).

Household industry is probably the most common Mesoamerican organizational level. Examples of this productive type are numerous, but Teotihuacán's obsidian-working sites are among the most studied and discussed (e.g. Clark 1986; Rattray 1987; Spence 1981, 1984, 1987, 1996). The bulk of obsidian production in this Classic urban center seems to have been carried out in residential compounds (Blanton et al. 1993; Cowgill 1997). Workshop industry is also represented at Teotihuacán. One of the best examples of this larger scale production level is a censer workshop with more than 20,000 pieces of molds and applications located in the North Quadrangle of the *Ciudadela* (Múnera and Sugiyama 1998; Sugiyama 1998). The village industry is perhaps best exemplified by a number of salt production sites in different regions in Mesoamerica. The Late Classic saltworks (AD 650-1000) at El Salado, in the Tuxtla Mountains is one such example (Santley 1994, 2004). Evidence from El Salado indicates a specialized salt making industry, represented by two major occupation zones: an area of high-density refuse accumulation where most of the salt was produced using the boiling method, and a lower-density area where the salt makers physically resided and some salt was made. The high-density workshop area was segregated from the habitation zone probably because the volume of that trash the salt making process produced was enormous (see Santley 1994, 2004). The nature of the metallurgical production at Itziparátzico might be better interpreted in the context of the larger Tarascan political and economic system.

Organization of metallurgical production: two models

Two alternative models of metallurgical production, which may provide a framework for interpreting the organization of the Tarascan copper industry, are presented below. Both cases are from colonial Africa, where important research on all aspects of metal production and use has been carried out since the early 1900s (see Childs and Killick 1993). For the purposes of this investigation, an emphasis is placed on ore processing and the production of metal rather than the manufacture of finished metal items. As explained in Chapter 3, the working of metals to finished products (e.g. decorative and ritual items, tools, weapons) is almost always a separate operation and does not apply to a primary production zone like Itziparátzico. In the Santa Clara region and elsewhere, the metals industry will be considered only as it relates to the processing of ore and the production of refined or unrefined metal in ingot form. Mining, as the actual process of raw material procurement, is also included in the discussion. The two models of metallurgical production presented in the following sections have been discussed by Raber (1984) and can be described as: 1) a local, small-scale industry based on part-time specialist labor and producing for an essentially local demand; and 2) a mobilized local industry, retaining many of the features of the local industry but subject to a greater or lesser degree of external interference or control and producing for a larger demand.

Organizational Model I: local metallurgical industry

The distinctive features of local copper production, according to Raber (1984), may be summarized as follows:

1. Production occurs on a small scale and is part-time or seasonal.
2. Production sites tend to be dispersed and small in size. Associated settlements are pre-existing agricultural (or other) villages and towns.
3. There is a small labor force composed of part-time specialists engaged primarily in other subsistence tasks, usually agriculture. The mining and smelting of metal ores is a seasonal occupation of some segment of the population.
4. There are few large or permanent facilities. A variety of techniques may be used in mining the ore but mining operations (pits, shafts, etc.) are few or small in scale and maintained by occasional local labor. Furnaces and other smelting facilities are temporary and constructed from locally available materials. New equipment and facilities may be constructed for each year's operations.
5. Storage and transportation facilities are provided by existing local systems.
6. Production is scaled to local consumption and scheduled according to local demand and the requirements of other (primarily agricultural tasks). It is therefore seasonal and timed to the agricultural off-season and seasonal weather patterns.

In terms of the organization of labor, the local metallurgical industry generally corresponds to Costin's independent specialization scheme. The scale of production, on the other hand, is closer to van der Leew's household industry category, although in some cases the workshop industry type may also apply.

Archaeological correlates

A local industry of the kind just described should leave distinctive archaeological traces that may include the following:

1. The presence of numerous small metallurgical sites located with respect to local outcrops of ore and within reasonable distance of local agricultural settlements. The sites should consist of many small and informal mine workings (pits and adits), dumps, and slag heaps representing opportunistic rather than planned exploitation of ores.
2. Little evidence of substantial or specialized architecture or metallurgical facilities. Furnaces and other features would be small and simple, reflecting occasional or temporary use. There may be little evidence of such features since old furnaces, for example, may be destroyed after use or incorporated into the next season's construction.
3. Evidence of a low level of production resulting from occasional smelting. There should be relatively small quantities of slag, gangue and other waste or byproducts.

4. Local consumption should be evident either in the local distribution of metal or in the absence of long-distance distribution.

An ethnohistorical example from south central Africa

Numerous examples of Colonial African copper and iron metallurgy may serve to illustrate the features of technological organization presented in Model I. Detailed individual ethnographies can be found in Cline (1937), which remains the most complete compilation of information on African metallurgy (Alpern 2005; Killick 1996; Raber 1984; Schmidt 1997; Schmidt and Avery 1978). For the purposes of this chapter, however, a single example of copper production has been selected: the BaYeke from the Katanga region in the Congo (Hemptinne 1926, cited in Cline 1937). Katanga copper was traded across the continent to Atlantic and Indian Ocean ports and both Arab and native markets for centuries. Living in an area of abundant and well-known copper sources, the BaYeke metallurgists have used local ores intensively, focusing mainly on the mining and smelting of copper carbonates, primarily malachite (see Chapter 3 for a reference to this type of ore).

The BaYeke smelters and coppersmiths were organized in guilds and their own initiation rites and tutelary spirits, with a very restricted membership. The right to hunt all animals except the elephant is also associated with this membership. This organization appears to be the result of the seasonal nature of metallurgy and the necessity for these part-time specialists to engage in off-season subsistence activities. Copper-smelting is a dry-season industry. Temporary furnaces (as many as 20 or 30 in a busy season) are constructed cyclically and destroyed after use. Each furnace requires minimum labor input, their construction from termite cones taking about half an hour. The furnaces are built near the malachite deposits and the smelting operations are carried out around the mines (Cline 1937). Technologically, these operations are fairly inefficient, as only about 50-60 percent of the copper present in the easily smelted carbonate ore is recovered. According to Cline (1937: 71), this technological inefficiency could be due to the absence of a suitable flux. It seems more likely, however, that this level of productivity is simply a reflection of both the abundance of ore and the relatively low demand (Raber 1984).

While the available ethnographic data do not provide details on the distribution and consumption of BaYeke copper, it seems clear that metal production is a seasonal part-time specialization scheduled around other economic activities. The length and intensity of the production period (indicated by the number of furnaces constructed, the weeks spent in smelting, etc.), appear to be subject to variations in local demand. The BaYeke example could be multiplied many times over across Africa and in countless other contexts. Although the specifics of technology and social arrangements vary, this local type of production shows a number of shared characteristics, including: 1) seasonal production; 2) scheduling; 3) part-time specialization (often an inherited status) by a small segment of the society; and 4) a low level of investment in labor, equipment and facilities.

Organizational Model II: mobilized local metallurgical industry

An alternative to the local industry just described is an industry in which production is mobilized for the purposes of a state or other central authority, or through a market system, but which remains essentially local in resource base, origin and scale. The salient features of such production are as follows:

1. This is in essence a local industry, based on the exploitation of local resources and originating in a local production system that has been incorporated or embedded in a larger system of demand, control, production and distribution.
2. Production may vary from part-time to full-time, according to the nature of demand and control by external factors (central political authority or market) and the degree of incorporation of local production into the larger economic system.
3. Such systems should include numerous dispersed and relatively small production centers located in relation to the distribution of metal ores, fuel and other resources.
4. The labor force, while skilled and specialized, may participate in production largely in response to the level of output demanded by external factors. The demands of state and market may require full-time production, but an industry of this type may also be consistent with part-time specialization and seasonal or periodic production.
5. The most distinctive feature of this form of production is its connection with a larger system of demand, control, production and distribution. The basic resources (ores, fuel, labor) are local, but since the industry is controlled by external forces (i.e. the state or the market), the products may be widely distributed through this system (Raber 1984).

The mobilized local metallurgical industry seems most consistent with certain forms of attached specialization in Costin's organizational scheme. The scale of production may encompass at least two of the categories proposed by van der Leew; the household industry and the workshop industry.

Archaeological correlates

The predicted archaeological features of an industry of the nature just described may be summarized as follows:

1. The regional pattern should be characterized by the presence of numerous small dispersed production sites, located with respect to access to resources. This pattern is similar to that of the local industry of Model I.
2. Archaeological evidence of metallurgical facilities should reflect the local nature of production. Existing settlements may provide the locations for such facilities. A relatively small local labor force will be largely supported by existing local arrangements. Thus, there will be no need for (or evidence of) of specialized structures or support facilities directly associated with the mines or smelting areas. Since the industry is essentially local, metallurgical facilities should be small in scale.
3. The amounts of waste, by-products and other metallurgical materials recovered in association with metallurgical sites should indicate a level of production above that of a purely local industry.
4. The pattern of exploitation (the extent of mining as indicated by pits and galleries, evidence of deforestation and other features) should indicate a planned, rather than an opportunistic approach to resource use.
5. Finally, the distribution of the metal produced in this industry may be quite broad.

The Kingdom of Benin: an ethnohistorical example from southwestern Nigeria

The organization of craft production in the Kingdom of Benin may provide an example of a Model II-type industry. Upon the arrival of the Portuguese in 1486, the Benin Kingdom was characterized by a high degree of centralization and segmentation (Bradbury 1957, 1967; Connah 1975; Roth 1968; Ryder 1969; see also Flannery 1972). The person of the king (Oba) and the capital city (Benin City or Edo) were the center of power and authority. Documentary evidence, however, indicates that the kingdom evolved from a more egalitarian base and local economy. By the 15th century, the Oba had become paramount within the kingdom, and the channel through which the spiritual world permeated the physical world. Through him, deceased ancestors continued to interact with the living. The Oba's power over the material world was just as great. He controlled trade and had the power of life and death over his subjects. His complete power was symbolized by the wall through Benin City that separated him from the local chiefs and townspeople (Bradbury 1957, 1967; Ryder 1969).

The Oba had direct control over almost all aspects of production. The spatial organization in Edo reflected this fact. The city was divided into two sectors: roughly one half was occupied by the palace compound and the other by forty heterogeneous wards, each of them consisting of one guild of craft specialists whose members traced a common ancestry. The most important crafts (e.g. bronze working) and their ward-guilds had representatives in the palace compound and some of their production was carried out there. While the Oba controlled all craft production, more exacting control was exercised over the manufacture of crafts essential to his own wealth and the validation of his ceremonial and political authority. Only the Oba could confer titles and display the ceremonial symbols of power. Certain items (e.g. bronze and brass goods) were demanded by the Oba for his exclusive consumption (Bradbury 1957, 1967; Connah 1975; Roth 1968; Ryder 1969).

The Kingdom of Benin was divided into fiefs led by officials, who were appointed by the Oba from the *uzama*, palace and town chiefs. The government was administered from the capital, and collected tribute from the villages. In addition to the city wards and guilds, some villages in the hinterland also specialized in a single craft and were tied to the capital and controlled by the Oba through craft guilds analogous to the ward-guilds of Edo (Bradbury 1957, 1967; Raber 1984; Ryder 1969). Although historical information is absent, it seems highly probable that the specialist villages were originally independent agricultural settlements with part-time specialists in certain crafts. The type of craft was probably related to particular local resources (clay, metal ores, etc.). As the process of political centralization intensified, demands for certain goods increased accordingly. These demands gradually became the demands of the central authority, materialized in the person of the king. Essentially local and part-time labor may in this manner, have been mobilized for the purposes of the Oba. With time, the Benin economy developed a high degree of control, exercised by a central authority for economic and political purposes (Raber 1984). The scenario just outlined, while admittedly speculative, represents a possible reconstruction of the process by which a local industry could be gradually incorporated into or integrated with a centralized political and economic system.

The organization of copper production in the Tarascan state: local versus mobilized industry

Empires can be classified as large, multi-ethnic states ruled from a single center (Sinopoli 1994, 2003). By definition, empires incorporate cultural and economic diversity. In the sphere of production, such diversity is manifest in scale and organization, and the nature and degree of integration of different productive areas into an Imperial order (Sinopoli and Morrison 1995).

By AD 1450, the Tarascan Kingdom had become the most important center of prehispanic metalworking in Mesoamerica. Metallurgy played a significant role in the structure of political and economic power in the Tarascan Empire. The state government, according to Pollard (1987: 745-746; 1993: 119), acquired finished metal goods and/or smelted ingots through different mechanisms, including 1) as gifts presented by foreign visitors and regional elite to the king; 2) by long distance merchants at the border of the Empire's territory; 3) as tribute paid to regional elites who in turn contributed part (or all) of the received goods to state storehouses in Tzintzuntzan; and 4) the direct movement of copper ingots from state controlled mines to the state storehouses.

The presence of craft producers with a considerable degree of specialization in the production of metal goods can be assumed from the relatively restricted spatial distribution of such products, the specific social and political context of their production and distribution, and the technological complexity of the industry behind their manufacture. Scattered references in documentary sources to these craft specialists support this assumption. While the consumption of finished metal goods was highly concentrated within a limited social and spatial territory, the production of metal from ore was naturally dispersed throughout the Tarascan territory (Pollard 1982: 258-259, 1987: 745). Ethnohistorical evidence suggests that production operations took place at a number of different locations within the domain of the Tarascan Empire (Pollard 1987: 748).

Ore deposits and mining

Apparently, the bulk of the metal that moved into the Pátzcuaro Basin came in the form of tribute delivered with regularity (see Paredes 1984; Pollard 1982, 1987). The primary supplier of copper was the central Balsas Basin, in the southern portion of the Tarascan territory, which is the region where the mining zones of La Huacana, Turicato and Sinagua are located (see Chapter 4, Figure 4.4). Paredes (1984) and Pollard (1987) have suggested that during the last century of the Tarascan Empire the state took more direct control of the copper resources of this particular region than simple tribute. This idea is largely based on accounts in the *Legajo* 1204 (Warren 1968; see also Chapter 3 in this volume), which suggest that the *Cazonci* (the paramount ruler of the Tarascan state) sent workers to extract copper from the mines of La Huacana to meet his needs (Pollard 1987: 748; Warren 1968: 47, 48). Some mines, however, continued to be exploited through the tribute system (Pollard 1987: 748). Pollard (1982: 258, 1987: 748) has pointed out that the schedule for payments of copper, of either every 40 days or on demand, substantially exceeds that of any other tributary item for the central authority.

The *Legajo* 1204 indicates that mining activities and smelting operations often took place at separate locations the central Balsas Basin. While mining centers were concentrated at Churumuco, Sinagua, Cutzian and La Huacana, among others, smelting operations were carried out at La Huacana, Cutzian and Huetamo-Cutzio (Pollard 1987; Warren 1968). At the Cutzian mine alone, there had been up to 50 miners and 40 others, some of whom were moving the dirt and mineral out of the mines and some were processing the ore (Warren 1968: 49). Accounts in the *Legajo* also declare that the metalworkers from La Huacana region owned and cultivated the fields at the foot of the hill where the copper veins were mined. This suggests that mining and metallurgy (at least at this location) represented part-time activities, undertaken mostly during the slack period in the agricultural season. The great climatic variation between the rainy and dry seasons in the region supports this assumption. During the rainy season, the mines were probably flooded, while during the dry season, agricultural production must have declined dramatically due to the extreme dryness in the region. The miners/smelters most likely alternated between metalworking and farming, according to the seasons as well as to royal demands (Grinberg 1996: 433).

Referring to the Churumuco region, the *Legajo* 1204 states that twenty smiths worked in each mine and each of them collected one-half *celemín* (Warren 1968: 37), that is 2.25 liters (Brand 1951: 132) of mineral, in the form of rock fragments and dust. After grinding the mineral and mixing it with ground charcoal, the metal-workers used blow-pipes to smelt it into an ingot one hand long, one hand wide, and two fingers high (Grinberg 1996: 433; Pollard 1987: 748; Warren 1968: 37). According to calculations by Grinberg (1996: 433) the twenty smiths together produced one *carga* (load) of copper per day, and one *montón* (charge) per month. One *carga* therefore was 20 ingots, and probably weighted about 90kg. A *montón* was the production achieved in 20 days, the length of a Mesoamerican month, which amounted to 400 ingots, weighting around 1800kg. These figures, however, seem too high and might require re-evaluation.

Smelting operations

Systematic research in the archaeological zone of Itziparátzico, the focus of the present study, has located potential production areas where concentrations of smelting slag were recorded. To recapitulate from Chapter 4, archaeological test excavations conducted at Itziparátzico produced large amounts of slag. The absence of metallurgical materials other than slag (e.g. crucible fragments, mould fragments, stock metal, metal prills, failed castings, part-manufactured objects and spillages, etc.) around Itziparátzico indicate that only primary copper production was carried out at the site (see Chapter 3 and 4). The analyses of slag samples corroborate that the production activities carried out at Itziparátzico involved primary smelting. The identification of un-reacted ore, together with prills of copper sulfide and copper, unambiguously characterize this material as copper smelting slag. The mineral ore processed was a sulfidic ore, chalcopyrite ($CuFeS_2$). The dominance of fayalite and the presence of copper prills indicate a high smelting temperature of around 1100°C, and reducing conditions (see Chapter 4; also Maldonado et al. 2005).

The most outstanding feature of Itziparátzico is its location about 125km away from the mines themselves. While this situation may not seem economically advantageous, the movement of ore from mine to settlement is well represented in the archaeological record. Several early examples from the Old World can be cited, including Chalcolithic sites such as Abu Matar (Gilead and Rosen 1992) and Shiqmim (Golden et al. 2001; Shalev and Northover 1987) in Israel, which lie some 150km from the nearest ore source and yet present substantial evidence of smelting activities. Excavation of the Chalcolithic settlement sites at Wadi Fidan in the Wadi Feinan area revealed slight but unambiguous evidence of smelting. In this case, however, the ore source is only about 2km from the smelters (Hauptmann 2003). More recently, evidence from the Early Bronze Age in Crete has indicated that smelting operations were being carried out at Chrysokamino, a site located in the northeast of the island, far away from any known ore sources. It has been suggested that the beneficiated ore was brought in by ship for smelting (see Betancourt 2006). Craddock (2001: 153) has suggested that copper was smelted at both mine and settlement sites form the inception of metallurgy, but processing methods have left little durable debris. The case of Itziparátzico is discussed in more detail below.

Independent or supervised smeltings?

Considering the apparent involvement of the Tarascan state in the extraction of ore, it is possible that the smelting operations were also under some degree of supervision. As the demands for copper of the central dynasty increased, one might suggest a model in which the Tarascans established a production and distribution system planned to obtain metal from different locations and divided the industry up into small components that could be easily managed: mining, transportation of copper (and probably tin) ores from different locations, smelting, alloying, finishing by casting and forging. The new system may have started with the smelting of ores away from the mines at sites like Itziparátzico, which had access to high quality charcoal (made of oak-wood, which has a far more efficient combustion than fuel made from scrub trees around the mines) for smelting,

and was located strategically on the route connecting the central Balsas region and the Pátzcuaro Basin.

The geographic situation of Itziparátzico is also important because of the potential use of the strong winds, which would provide a constant draft of air for highly efficient smelting processes (see Chapter 6). The increasing demands of the state for copper may have favored the introduction of such technological innovation. Ore had to be carried some 125km, which would also imply an increase in efficiency of the transport system to cope with high transportation costs. According to accounts in the *Relación de Michoacán* and the *Legajo* 1204, processed ingots were transported for final manufacture on the backs of *tamemes* (the Nahuatl term for human carriers), each load consisting of 20 to 30 ingots (Warren 1968: 47, 49) which weighted 32-72kg (Pollard 1987: 748). Based on information found in the *Legajo*, Pollard (1987: 750) has estimated that the major smelting centers of La Huacana and Cutzian were two days' journey from Tzintzuntzan. These figures make the transport of ore to a relatively intermediate point like Itziparátzico seem plausible.

Itziparátzico does not represent an isolated occurrence in the Tarascan territory. Large deposits of slag in (presumably) prehispanic contexts have been recorded at Jicalán El Viejo, a site that occupied a strategic position on the frontier between the *Tierra Caliente* (to the south) and the Sierra Tarasca (to the north) (Roskamp et al. 2003) (see Chapter 4, Figure 4.8). Slag samples from Jicalán were subjected to the same analyses as the samples from Itziparátzico, producing similar results: evidence of a smelting technology that used sulfidic ores and very efficient furnaces. Incidentally, the site is also located in an area where wind-powered pyrometallurgy seems feasible (Roskamp, personal communication). Both the organization of production and the complexity implied in the task of smelting sulfidic ores seem to have involved full-time craft specialists. Conversely, if consistent with mining activities (and the wind regime), the smelting operations may have been seasonal. Evidence for the presence of specialized Tarascan smelters is found in a testimony in the *Legajo* 1204, which in reference to copper smelters in La Huacana region in 1533, states that

> que los indios comunmente no lo saben soplar ni hacer [el cobre] sino son los que son maestros que en cada pueblo donde hay minas hay doce o quince fundidores que lo saben sacar (Warren 1968: 37).

> the indians do not commonly know how to heat up or make [the copper] but only the ones who are masters [know]; in every town where there are mines there are twelve or fifteen smelters who know how to obtain it.

Final processing and manufacturing activities

The types of metal artifacts and manufacturing methods and materials (pure metals and alloys) that Tarascan metalworkers employed in their fabrication have been relatively well documented (e.g. Grinberg 1990, 1996, 2004; Hosler 1988a, 1988b, 1988c, 1994; Pendergast 1962; Rubín de la Borbolla 1944; see also Chapter 3). Copper was alloyed with tin and/or arsenic to produce bronzes, or mixed with various concentrations of silver, gold, or both, to fashion bells, ornamental tweezers, rings and body ornaments. Although bronzes were also used to manufacture tools such as axes, hoes, and needles, the main focus of Tarascan metallurgy was on sumptuary objects that reflected fundamental religious and political agendas. Metals were directly associated with particular deities and the *Cazonci*, as the human representative on earth of the patron god *Curicaueri*, also shared this association (Hosler 1994; Pollard 1987). Given these facts, it is not surprising that the greater part of the metal items produced in the Tarascan territory were concentrated in the royal palace and associated quarters of the social and political elite in Tzintzuntzan, and in royal treasuries on the Lake Pátzcuaro Basin. Outside the core of the empire, metal objects were found in relatively small quantities in elite contexts and, to a limited extent, as implements for basic production (Pollard 1987: 745).

The *Relación de Michoacán* makes reference to a kind of 'guild' system in the crafts in the Tarascan core, each craft being under the authority and supervision of a subject of the *Cazonci* (Martínez S. 1903: 171-178; Warren 1985: 20-21). While the presence of guilds is questionable (since no evidence of their existence has been reported anywhere else in Mesoamerica), apparently artisans who were in some way attached to the palace in Tzintzuntzan produced a wide range of goods for the royal household, including objects of gold, silver and copper. It is not clear whether these were royal retainers or individuals paying tribute in specialized activities (Pollard 1982: 259; Gorenstein and Pollard 1983: 103), the evidence however, points toward a category closer to the former.

As discussed earlier in this chapter, attached specialization (following Brumfiel and Earle 1987, and Costin 1991) has been conceptualized as one end of a continuum (opposing independent specialization) in the relationships of producers to sociopolitical institutions and/or patrons, in which institutions or elites directly regulate some aspects of craft production, including control over raw materials, control over access to finished products, and control over producers themselves, among others. In this case the concept applies, because the raw materials (metal ingots) were owned by the state, the distribution of metal goods was socially and spatially restricted, and the specialists in the craft worked under supervision. Given that the manufacture of metal objects was also taking place outside the capital, it is likely that analogous contexts of attached craft production existed in elite compounds in the administrative centers of the Pátzcuaro Basin.

The storage of ingots brought from the smelting sites ensured a stable supply of raw material for metalworking, which may have also facilitated production of metal goods as the primary economic task the craftsmen were engaged in, that is, full-time specialization. Although no references have been found of the principles for the recruitment of these specialists, it is well known that in Aztec society masters of trades taught their profession to their sons from childhood (see Berdan and Anawalt 1992: 145, for an illustrated example). It is possible that the system in the Tarascan state in the Late Postclassic period was similar to that of Central Mexico at the time. Nahuatl speakers also used two different terms to distinguish between 'gold casters' and 'copper finishers' (Sahagún 1969-1982: Book 10, part 11, ch. 7, figs. 33, 34). The *Diccionario Grande de la Lengua de Michoacán* (Warren 1991) makes a clear distinction between the terms metal melting (*tiyamu ytsieranstani*) and metal casting (*yurehpanstani tiyamu*), which may indicate a comparable subdivision within the metal craft. However, this separation of terms could have served simply to describe different stages of manufacture. While we do not know the nature and size of the production units, both the patterns of distribution and the context of production suggest small workshops, probably inside the elite households. Given the apparent spatial circumscription of metalworking, the complete process of production of an artifact, including melting, alloying and casting or forging, may have been in the hands of the same individual or individuals.

Discussion

The diverse arenas of copper production in the Tarascan Empire present very different pictures in terms of the forms and degrees of craft specialization and elite involvement and control. Here, like in other contexts of production and consumption of specialist craft products, individuals and groups are linked in intricate webs of interdependence and interaction. Tarascan copper production involved social groups of various sizes, including raw material procurers or producers (who may include other specialists, e.g. miners, charcoal producers, and smelters who provided ores and smelted materials to forgers or casters), to people involved in various stages of production and distribution. Producers also interacted with those who acquired and used their products (i.e. elite patrons). The scale of production seems to have varied accordingly, ranging from household industry and workshop industry to village industry.

In terms of the general organization of metallurgical production, the Tarascan case seems consistent with a mobilized local metallurgical industry, although in some specific instances the model does not fit the data adequately. The evidence from Itziparátzico clearly indicates a non-local industry, since the ore and metal ingots had to be transported long distances. In this case, an adaptation of the model can be suggested which accounts for this variation; this would be a mobilized regional metallurgical industry. It seems clear, however, that some mining and smelting of copper ore and goods manufacturing was also taking place outside the main system of the state and its centralized control, possibly in household contexts, as indicated by the evidence of tools and items of daily use. This additional dimension of production opens up a whole range of variation to the Tarascan copper industry, and would conform better to the local metallurgical industry model.

The the multi-dimensional Tarascan production model presented above, while fragmentary and requiring further data collection and studies, clearly reflects the complexity of this dynamic system for a single craft. Its main organizational characteristic was the division of an industry into small segments that could be easily controlled, including mining, smelting, and final processing (and probably a range of subdivisions within these). With no single section performing all the needed manufacturing steps, centralized control was more easily established and maintained. Control, however, seems to have been exercised in a variety of ways with varying impacts and consequences. The political ecology view applied here, which emphasizes the dynamic between centralized control over production and exchange, on the one hand, and dispersed strategies for managing different localized or dispersed economic resources on the other, is therefore the most suitable approach to explore this organizational structure.

The distinction between attached and independent specialization understood as a continuum of variability, as formulated by Costin (1991), can be used (although not without caution) to explore the nature of the relation between the different categories of producers in the copper industry and the Tarascan state. In this context, such a scheme is perceived as a set of particular relations of production with specific duration, rather than fixed and exclusive relations of production (see Smith 2004: 83; Sinopoli 2003: 32-33; Stein 1996: 25-26). Attached specialists are generally viewed as being dependent upon, and bound to, institutions, while independent specialists exercise greater choice over their activities and labor. As pointed out earlier, however, considerable variability can lie within each of these relations. Copper production in the Tarascan Empire appears to have encompassed, for at least a period of time, different forms of variation in the production units.

Mining in the Tarascan territory was regulated mainly through the tribute system, although it also involved some direct exploitation by the Tzintzuntzan elite (see Paredes 1984; Pollard 1982, 1987). These varying state procurement strategies may have resulted in units of production units that correspond roughly to at least two of the categories suggested by Costin (see 'Costin's gradients of variation' above). One of these productive types is dispersed corvée production (e.g. labor in mines and smelters to produce tribute for the state), and the other one nucleated corvée (e.g. workers sent to state mines). Smelting operations are the least known and most problematic aspect of Tarascan copper production. It is likely that the degree of control by the state varied by region, in a similar way as mining. The presence of elite items at a production site (Itziparátzico), however, may imply a sort of individual retainer production (e.g. individual skilled artisans recruited to produce for elite patrons or state institutions) taking place in particular locations. The fact that the land in this region (as in the rest of the Zirahuén Basin) seems to have been in the hands of the Tarascan nobility (see Chapter 4) supports this assumption.

Finally, artifact manufacture is consistent with forms of retainer production in which metalworkers produced valued goods for the elite and the palace in supervised units. Metal ingots were presumably kept in state storage houses. This suggests that control of the final processing may have been exercised by restricting access to the raw materials for metalworking, rather than by controlling specialized labor. Casting, forging, and finishing, while labor intensive, are relatively easy tasks once the basic skills have been mastered. During the preliminary stage of the present research, I worked on the experimental replication of copper and bronze artifacts and was able to produce several axes, Tarascan tweezers, and lost-wax cast bells (Figures 3.8, 3.10 and 3.11a, b in Chapter 3; see also Maldonado 2005). I had no previous knowledge of the materials, tools and techniques involved in the making of these artifacts. Nevertheless, with the instruction and supervision of a full-time artisan and through trial and error, I successfully replicated a total of thirteen pieces in a few weeks.

If the model above applies, control from the elites and the central power in the Tarascan state was aimed mainly at limiting access to both raw materials and finished products. By extension, the craft producers themselves may also have been under their control. Sumptuary laws were probably also used to regulate the display of metal objects. From the standpoint of a political ecological approach, two possible scenarios emerge. One scenario suggests that institutional representatives may have directly and forcefully imposed control, or exerted their 'power over', the production process (as did the Oba in the Kingdom of Benin). In the alternative scenario, control of informed actors may have been more subtle and indirect, involving inducements and rewards rather than coercion, that is, utilizing their 'power to' control craft production (as did Aztec rulers who patronized certain crafts; see e.g. Blanton and Feinman 1984). The dialectic of control (or the relation between agency and power; see Chapter 2) lies between these two scenarios.

This chapter presented an attempt to link the technological processes identified at the archaeological zone of Itziparátzico to their wider cultural, social and economic context in the Tarascan Empire. Two general models of the organization of production were introduced as alternative explanations for the organization of copper production in the Tarascan state: the local metallurgical industry and the mobilized local metallurgical industry. Both models seem to apply to Tarascan copper technology in different spheres and levels of the productive system, and both are characterized by production on a small scale as a seasonal activity by villagers engaged primarily in agriculture rather than by full-time specialists. A variant of the mobilized industry model was considered, in which the regional scale of the resource procurement and processing is taken into account. While emphasis was on ore procurement and extractive metallurgy, final processing was briefly discussed as an important part of the general *chaîne opératoire* for Tarascan copper production. This last aspect of the operational sequence involves a more direct intervention of the central power in the transformation process.

The present study, while focusing on a specific craft, could not isolate it completely from the larger Imperial system, due to the multi-scalar and interrelated nature of the social units of production. The above discussion illustrates some of the political, social, economic and geographic variables that influenced the production of copper in the Tarascan state. The kinds of social and economic relations that developed among producers and between producers and consumers seem to have varied considerably throughout the empire (and probably over time). The nature of these relations and the reasons for their patterning are important questions to address in future research. The exploration of these relationships will enable us to consider both specific cases and broader cross-cultural patterns. Prospects for future research will be discussed in the next section of this volume.

Chapter 6

Conclusions, final remarks and suggestions for future research

The introduction to this volume highlighted the limited knowledge and understanding of the procedures of extractive metallurgy in western Mexico, as well as its social context, identifying it as one of the major problems in comprehending Mesoamerican technology. The work herein presented contributes to resolving of these outstanding matters through the formulation of specific research questions and the generation of solid data. Due largely to budget constraints, however, these objectives have been only partially attained. It is paradoxical that this study should come full circle to arrive back almost where it began, with more questions than answers regarding the prehispanic western Mexican metallurgical industry. Nevertheless, along the way a new set of evidence has been collected. Although the subject remains the same, the point from which it is viewed has changed. The fieldwork of this study was carried out in two stages. The first objective was to reach a broad understanding of the complete archaeometallurgical record of Itziparátzico, the scale and nature of metallurgical practices of the past, and their cultural and chronological settings. This was achieved mainly through archaeological and ethno-metallurgical surveys, although other sources of information, including ethnohistorical documents were also assayed. From this overview, a number of questions concerning specific aspects of the record were raised; the most important of these related to the evidence from smelting activities in the area. Those questions were addressed during the second stage of the study, which focused on resolving the nature of the technology represented at Itziparátzico.

The strategy adopted during the second stage of the investigations was a test-pitting program, with three sectors of the archaeological zone targeted for investigation. This stage added a substantial body of evidence and confirmed many of the features observed during the survey. However, the data presented something of a dilemma in that they did not fit preconceived notions of Mesoamerican metallurgy. In essence, the epitome of prehispanic smelting is embodied in the crucible-bound and mouth-blown model. Notwithstanding, the presence of tapping slag (normally an indicator of furnace technology) at Itziparátzico contradicts the accepted wisdom, pointing toward other technological possibilities (i.e. the use of some form of natural draft).

Finally, the laboratory analyses strongly support the views derived from fieldwork data that smelting processes were carried out in furnaces designed and built to provide reducing conditions during the refinement of copper ores. This study has produced the first comprehensive analysis of slag from a Mesoamerican context. Samples from the presumably contemporary Jicalán El Viejo were also analyzed for comparative purposes using the same methodology. The results obtained indicate that both the ores and the smelting processes used were similar to those identified at Itziparátzico. Slag findings from other sites in Michoacán (Grinberg 1996) and from El Manchón, in Guerrero (Hosler 2002), as well as from more remote locations such as Colonial Lamanai in the Maya area (Simmons 2005), have recently been reported, and a new set of analytical data may soon be available for further comparison.

Evaluating the data from Itziparátzico

Two major problems derived from the paucity of data on Tarascan extractive metallurgical processes were outlined in the introduction to this work. First, the knowledge of the technology itself is fragmentary. Second, nothing is known about its associated resource control and contexts of production. The aim of this project was to examine the technology and organization of the copper industry at a Tarascan locale (Itziparátzico), relating the evidence for the size and organization of the industry to the character of technological processes, resource requirements, demand, and potential and actual production levels. While a number of questions remain unanswered, the present study has made substantial steps toward fulfilling the initial goals of the project, and several contributions have originated from research undertaken for this study. These contributions are best summarized as they relate to the four specific issues addressed in Chapter 1: theory of technology, archaeometallurgy, archaeometallurgical materials, and the organization of copper metallurgy.

Theory of technology

As pointed out earlier, understanding Tarascan metallurgy and its social context entails not only the acquisition of new information, but also the use of an adequate theoretical framework that encompasses both the technical and cultural aspects of the industry. This study required organizing a multi-approach perspective to address the specific research questions and problems concerning the copper industry at Itziparátzico. The most valuable theoretical lesson derived from this work is that because technology is a complex, integral and active component of dynamic human societies, generalized models to examine technological phenomena are more suitable for specific systems when combined with other complementary schemes.

The notion of scale was critical in the construction of this framework, since the goal was to develop an understanding of large-scale patterns and processes and their relationships to patterns and processes at smaller scales. The combination of approaches advocated in this study of Mesoamerican metallurgy includes large-scale ecological and evolutionary models, as well as small-scale behavioral and processual schemes. This sets out from the premises that behavior is a mediator between social and physical opportunities and constraints and technological production, and that people use diverse strategies to achieve goals or solve problems.

Departing from this deductive-inductive approach, Tarascan metallurgy was first contextualized within the broader frameworks of Mesoamerican technology and craft production. The Mesoamerican system was linked by continuity in the use of particular materials as items of wealth and standards of value, and this was reflected in economic, social, and political practices. Craft specialists worked fine ceramics, metals, obsidian, jade and other greenstones, and feathers into signs of distinction; scribes, astronomers and calendar specialists developed and recorded indigenous wisdom; and a select body of people who claimed legitimacy in exercising powers of governance consumed these and other forms of 'high culture' (Joyce 2000). The Tarascan metal industry was part of this larger system.

The implementation of such a dynamic theoretical construct was also important in attempting to get away from technological determinism and look at the concept of metal technology as a social phenomenon. Metallurgy in Mesoamerica followed its own trajectory of innovations and developments, which could be reflections of social organization or the structure and worldview of local communities, sometimes coupled with environmental factors within that region.

Archaeometallurgy

At a more particular level, the perspective used in this study, in combination with multiple lines of evidence, allowed me to hypothetically reconstruct the *chaîne opératoire* for Tarascan copper production. As expressed elsewhere in this work, due to the paucity of data and limitations of available analyses, the archaeometallurgical record for Mesoamerica is fragmentary and dispersed. Most of the available information on metallurgical processes is largely based on metallographic analyses of finished products and therefore often restricted to the final stages of production (i.e. fabrication, surface treatment and finishing of metalwork).

This reconstruction nevertheless was made possible through the use of multiple data sources, including archaeological and ethnohistorical evidence, conceptual, scientific and technological analyses, and cross-cultural comparisons. In studying the operational sequence of producing metal artifacts from copper ore, the interplay between technology and organizational forms could be observed from the procurement of the raw material (copper) to the achievement of the final result: the metal object. In the process of reconstructing this metallurgical sequence, fundamental technical terms and concepts associated with preindustrial metallurgy were introduced. This conceptual and technical review represents the first steps toward the creation of a much needed archaeometallurgical glossary for Mesoamerican metallurgy.

Archaeometallurgical materials

Although, unfortunately, no identifiable metalworking structures (furnaces, hearths, and pits) were recorded during the investigations at Itziparátzico, substantial evidence of copper smelting (i.e. several hundreds kilograms of slag) was collected. The basis of metal smelting is to take a metal rich ore and chemically reduce it at high temperatures, leading to the separation of a metallic phase and a slag phase. The slag is formed of all the components of the ore/furnace system that are rejected by the metal. Thus the metallic phase will consist almost exclusively of the primary metal, leaving the slag with a virtual encyclopedia of chemical information about the smelting process in the form of silicates. The slag is, therefore, a liquid record of each individual metal producing smelt that is then rejected as waste.

Slag has a pivotal importance in the field of archaeometallurgy. Notwithstanding, no comprehensive studies of slags from Mesoamerican contexts existed prior to my research. The present work thus represents the first contribution toward the creation of a slag database for Tarascan metallurgy.

The results of the analyses of slags from Itziparátzico indicate a smelting technology that used sulfidic ores and highly efficient furnaces. While further archaeological investigations are required to precisely date these activities, this technological information is important for establishing the context and scale of copper production at the site. These results have set the stage for further research both in the Itziparátzico area and at the regional level.

The organization of copper metallurgy

The main objective of the present study was to construct a model for Tarascan copper production based on the findings of the research described above, in order to evaluate the assumption that mining and metallurgy were under the control of a central power. This objective has been achieved, once again, through the use of multiple lines of evidence. Two generalized models of organization of metals industries derived from ethnographic information were considered for comparison: 1) a local, small-scale industry based on part-time specialist labor and producing for an essentially local demand; and 2) a mobilized local industry, retaining many of the features of the local industry but subject to a greater or lesser degree of external interference or control and producing for a larger demand. The nature of local production in the region was related to larger economic and political patterns and the scale and nature of demand and organization of the industry were all found to be subject to the influence of external economic and political factors.

It became apparent during the construction of this paradigm, that the copper industry could not be isolated from the larger economic system of which it formed a part. This is due to the multi-scalar and interrelated nature of the social units of production. The resulting model infers some of the political, social, economic and geographic variables that influenced the copper production in the Tarascan state. The kinds of social and economic relations that developed among producers and between producers and consumers seem to have varied considerably throughout the empire, and probably through time. The nature of these relations and the reasons for their patterning are important questions to address in future research.

During the course of this investigation, a number of topics for further research were identified. These avenues of research were beyond the scope of the current study. However, if funding is available in the future, investigation of the following issues may provide valuable information.

Potential for future research in and around Itziparátzico

The Itziparátzico Archaeological Research Project 2003-4 represents the first systematical study of a metallurgical zone in the Zirahuén Basin, and one of very few in Mesoamerica. Needless to say, the most critical next step is to undertake further investigations in the area and its surroundings. Although survey and test-pitting were relatively successful in revealing the chronology and character of the archaeological zone of Itziparátzico, they also raised a number of issues that could not be resolved without further research. An in-depth analysis of ceramic and lithic artifacts is needed to more precisely establish the chronological and cultural relationship between Itziparátzico and other areas, particularly the Pátzcuaro Basin.

A more extensive survey of the area will begin to assess the nature and extent of the settlement (or settlements). Additional excavations would be required to attempt to date the materials and features more accurately and also to explore more fully the sector that contains most of the metallurgical evidence (Sector 3). Systematic quantification of slag is needed to obtain estimates of copper production at Itziparátzico. Such estimates can be made by calculating the copper yield; that is, the percent of copper in the ore that was extracted by the smelt. Slag analysis has determined that the ore being processed was chalcopyrite ($CuFeS_2$). Because the chemical composition of the slag reflects that of the ore, and since the chemical composition of this mineral is known (roughly 30.43% iron, 34.63%, copper, 34.94% sulfur), a good estimate of the copper yield may be obtained from the total weight of the slag. Finally, only the recovery and analysis of furnace remains can resolve the most pressing question on the smelting processes carried out at Itziparátzico: were the furnaces used bellows driven or wind powered? The answer to this question will have important chronological and technological implications.

Technological and chronological implications

Analyses of slag samples revealed that the technology employed in the smelting works at Itziparátzico involved an efficient, highly reducing smelting environment, which would be difficult to achieve using lung-powered blowpipes (see Rehder 1994). As pointed out in Chapter 4, if the smelting activities at Itziparátzico involved the use of bellows, we may be dealing with a relatively early post-Contact production area which largely overlapped with prehispanic traditions, as indicated by the evidence of Tarascan pottery and lithics. Conversely, if the operations at the site are Late Postclassic in date, then it is possible that Tarascan metalworkers had developed a copper smelting method that involved some form of natural draft. The exposed position to prevailing strong winds during the dry season in this area seems to support the idea that a naturally wind-aided processing may have been taking place at the site. It is important to mention that no evidence has yet been found to prove the use of wind power in Mesoamerica. Nevertheless, little is known about Mesoamerican smelting technology and further research should explore this possibility. The use of wind power for smelting metals, however, has been suggested for a number of sites in both the Old and New Worlds.

Comparative models for the use of wind power in preindustrial smelting

Because of the limitations of the archaeological project at Itziparátzico, no systematic efforts were made to investigate the potential for the use of wind power for copper smelting in the area. Substantial research will be required before an acceptable set of indicators for this type of metallurgical technology can be developed. The most critical element in such research is the location and investigation of smelting furnaces. Further archaeological research needs to be conducted to locate furnaces and other production features. Without this key piece of information no steps can be taken to scientifically assess the potential for wind use in the area.

The results of fieldwork research and slag analyses, however, provide a solid enough basis to consider this technological possibility. A model derived from Old World archaeometallurgy may provide the framework for designing future research in western Mexico. Wind-power systems are not exclusive to the copper industry. Gillian Juleff (1996, 1998a, 1998b) has carried out extensive work on the use of wind-pressure systems for iron and steel production from the first millennium AD in Sri Lanka. Her data from both survey and excavation in the Samanalawewa area, in the Upper Walawe basin (central highlands) pointed to the utilization of furnaces powered by wind.

According to Juleff (1998a), neither conventional archaeometallurgical nor technological knowledge could not provide an adequate explanation for the findings at Samanalawewa. On the contrary, these data appeared to deviate from the accepted model for early metallurgical processes, based on a general classification of furnace types which recognized a basic progression from 'less advanced' bowl furnaces to 'more advanced' shaft furnaces (Juleff 1998a). Juleff and her team built replica furnaces based on the remains of those found in archaeological contexts.

Smelting trials revealed that monsoon winds blowing over the tops of the furnaces' front walls caused air to be drawn through conduits into the furnaces at a continuous rate. About half of the metal produced in these trials was high-quality steel. Juleff (1998a) has also pointed out that in many regions in the world the conditions exist for the use of wind-powered furnaces. These findings have contributed significantly to the growing evidence of the use of wind power in preindustrial metallurgy. Because of the relevance of its methodology as a reference for future research in the Tarascan area, this Sri Lankan model is described in more detail in Appendix E.

A brief overview of wind-powered smelting technology

For numerous reasons, the concept of a wind-driven furnace had until recently been an anathema in archaeometallurgy. The basic contention was that no wind is free of gusting and therefore could not provide a dependable source of energy, particularly over the relatively long periods of time (hours) that most smelting operations require for successful completion. In addition, wind appears to offer no opportunity for controlled air supply. The tacit assumption was that to be effective, the wind would have to emulate the action of bellows by blowing directly into the furnace and also that under non-constant conditions a smelting process would fail, rapidly and irretrievably (Juleff 1998a). The work of a number of scholars and particularly the pioneer research of Juleff in Sri Lanka, have changed this perspective considerably (see Appendix E). Indeed it appears that Early Bronze Age copper smelters in the Middle East and the Eastern Mediterranean had recognized and harnessed the potential of the wind for smelting meta (Craddock 2001; Juleff 1998a).

The transition between the Chalcolithic and Early Bronze Ages in the Middle East and Eastern Mediterranean (around 3000 BC) is characterized by evidence of the earliest recorded use of furnaces specifically built for metallurgical processes. Apparently, such furnaces were designed and located to take full advantage of prevailing local winds (Craddock 2001). The most comprehensively investigated sites are in the Wadi Feinan on the east side of the Wadi Arabah, in southern Jordan (Hauptmann 2003). Remains of small furnaces positioned at prominent places close to the tip of mountain slopes, were found under scatters of slag very carefully set into the prevailing winds. Smelting experiments carried out by Hauptmann and colleagues (see Kölschbach et al. 2000) confirmed that wind-powered furnaces were suitable for producing copper and liquid slag. Similar finds were made on a hilltop at Timna, in the Western Arabah (Rothenberg and Shaw 1990).

Evidence of wind-blown furnaces has been recovered from the area between the Eastern Desert and the Gulf of Suez, in Egypt. The furnaces were located just below the crest of step ridges and faced into the prevailing winds. Slag analysis indicated furnace temperatures of around 1150°C. Recent discoveries have also been made in Crete, at the site of Chrysokamino, in the northeast of the island. The furnaces were sited in a small gap between two peaks near the sea to take advantage of the wind. Chrysokamino is also remarkable because of its location far away from known copper sources, which suggests that (as in the case of Itziparátzico) beneficiated ore was transported long distances for smelting (see Betancourt 2006).

In pre-Columbian South America, wind-drafted furnaces lined with clay, or *guayras* (also known as *huayras* or *huayrachinas*; from Quechua, *wayra*: air, wind) were used to smelt metal ores. Some *guayras* were made of rough stones, loosely assembled so that the wind could blow through the gaps and fan the fuel. The more complex ones were built of stones set in clay where holes were left open to allow the wind to enter (Bakewell 1984; Peele 1893). Portable clay *guayras* were developed after the Spanish Conquest, designed specifically for extraction of silver (Bakewell 1984: 15; Capoche 1959: 110). Typically, all *guayras* were set on exposed ridges where the wind blew strongly (Bakewell 1984; Petersen 1970). Pedro Cieza de León was struck by the efficiency of *guayras* when he visited Potosí in southern Bolivia in 1549, particularly because Castilian style furnaces with bellows would not function there (Cieza de León 1985: 449, ch. cix). By the 1580s almost all *guayras* were abandoned and the method of amalgamation (the blending of pulverized ore with mercury) became the dominant technology of refining, and so remained for the rest of the Colonial period and beyond (Bakewell 1984). Guayra technology has been investigated in conjunction with an ongoing multidisciplinary project to study prehispanic and historic period silver mining at Porco, a mining center located 50km southwest of Potosí (Van Buren et al. 2006).

Other pressing research needs: fuel production studies and ethnohistory

Several important areas of research concerning the archaeology and archaeometallurgy of Itziparátzico and its surroundings remain to be investigated. One of these is fuel production. The production of charcoal fuel from local timber appears to have been a key factor in the establishment of the smelters in the area. Natural regeneration of forest communities probably kept pace with production requirements. Anthropogenic deforestation for charcoal fuel appears minimal in the region even at the height of Colonial production in the eighteenth century (see Davies et al. 2004; Davies et al. 2005; Endfield and O'Hara 1999). Similar situations have been observed for other areas in the world, including extensive tracts of land in Africa, Europe and the Middle East (Juleff 1998a).

References to charcoal making (*turiri curirani*, the native term in Tarascan) and charcoal makers (*turiri curirati*) are frequent in ethnohistorical documents (e.g. Warren 1991). Charcoal is still produced in the Santa Clara area using traditional techniques. At present most of the fuel used by the Santa Clara coppersmiths is made in the neighboring town of Opopeo. Unfortunately, no systematic research on the production of charcoal has been carried out in the Basin of Zirahuén. Prehispanic charcoal making probably existed as an industry separate from metallurgy. Charcoal, however, was the main fuel and reducing agent used in furnaces for smelting copper. A study of charcoal production would provide us with a more accurate representation of the scope of the smelting operations in and around Itziparátzico.

A second critical area needing research relates to the ethnohistory of Itziparátzico and its surroundings. Spanish accounts make little mention of prehispanic settlements in the Zirahuén Basin. Data from the five known fragments of the Carvajal visit to Michoacán (in Warren 1985: 58), mention a village known as Icheparataco, which was subject to Erongarícuaro (Eronguariquaro), one of four major administrative centers under Tzintzuntzan's control. Although Icheparataco seems to be a distorted version of Itziparátzico, Helen Pollard (personal communication, 2003) has pointed out that Icheparataco was probably located about ½ league (2km) from Erongarícuaro. Itziparátzico, on the other hand, is 25km SW of Erongarícuaro (see Figure 4.6). According to Pollard, it would be unusual for a community located south of Pátzcuaro (a separate administrative center) to be subject to Erongarícuaro on the SW portion of the basin. Only close scrutiny of the original sources will help us positively identify Itziparátzico and other similar communities in their physical and political context. Establishing the status of these communities within the Tarascan domain at the time of contact will be key to determining their role as metal producers in the economic organization of the state. A search for ethnohistorical data on the region in colonial archives is therefore a priority.

Potential for future research at the regional level

Earlier in this volume, the issue of scaling in technological and craft production studies was put forward (see Chapter 2). As has become evident during the present study, small-scale behavioral and processual phenomena cannot be separated from large-scale ecological and evolutionary processes. This scaling principle also applies to local developments in connection with regional systems. The evidence at Itziparátzico has to be approached at both levels, in its specificity, but also as part of a larger socio-technological context in both the Zirahuén Basin and the broader Tarascan region. An adequate understanding of the organization and development of metallurgy or any other technology rests on an understanding of the number of metallurgical sites or features, the resource base, and contemporary settlement patterns, economics, and political organization. Without this information, the study of individual metallurgical sites in western Mexico will require the correlation of several incomplete lines of evidence and remain tentative at best.

The archaelogical study of metallurgy or any other early technology must involve a consideration of the environmental factors affecting the supply of raw materials. In this study, inferences had to be made about the fuel supply at Itziparátzico, the nature and distribution of ores used by prehispanic metallurgists, and general environmental conditions in the Tarascan territory. Systematic archaeological research must be conducted at locations in the mining district in the Balsas area to fill in critical gaps in our knowledge of the prehispanic production of copper in the Tarascan territory. The study of the patterns of exploitation of wood for charcoal making at the regional level, for example, may reveal valuable information about both settlement patterns and technology. Our archaeological knowledge of settlement, political organization and economics in the Tarascan state is also restricted.

Technological developments cannot be linked to cultural processes unless we have an adequate knowledge of changing settlement patterns, social organization, demography, etc. In the absence of archaeological data on these topics in the Zirahuén Basin, the present study has relied on inferences drawn largely from analogous situations elsewhere in the Tarascan region, Mesoamerica, and even other areas of the world, in considering the organization of metallurgy. Patterns of production, distribution and consumption can only be examined when appropriate information is available on regional settlement patterns, population size, distribution, and character, the existence and importance of markets and other economic factors, class structure, and political institutions.

On the basis of the retrospective knowledge outlined above, the following suggestions for future research may be made:

First, archaeological research on technologies should, whenever possible, be coordinated with regional archaeological projects pursuing broad anthropological and historical concerns. It is likely that the study area boundaries can be defined similarly for the study of both technology and other aspects of culture. The Tarascan region and other areas in Mesoamerica provide nearly ideal situations in which to investigate the development of technology and other aspects of culture, such as political economy. Cooperation in the study of these subjects should be encouraged and planned.

Second, attention should be directed to local pottery sequences as a basis for creating a regional ceramic sequence. This is particularly relevant for the evidence at Itziparátzico, since dating of slags and other metallurgical artifacts from the Late Postclassic Period in the Tarascan region can only rely on relative chronological methods. Comparative studies of metallurgical features from contemporary sites may also contribute to the understanding of Tarascan metallurgical (and Mesoamerican) technologies.

Third, complete and reliable environmental information is needed including paleobotanical data and information on climate, geomorphology, and ore geology and chemistry. Without such information it will be impossible to adequately evaluate resource (e.g. fuel and ore) availability, the factors affecting settlement patterns, regional limitations on population and production, and technological organization. Specifically, there is a need for a comprehensive program to test and characterize the ores available to native miners and metallurgists. Both trace and minor element analyses of ores are needed to enhance our understanding of the sources and amounts of ores available. The extent and rate of deforestation in certain areas could also be determined in order to estimate potential fuel supplies.

Fourth, chemical analytical data on metallurgical products and by-products must be obtained and used in association with specific questions on the character of prehispanic metallurgy. These issues must be phrased in terms of broad anthropological and historical concerns. Questions on technique or the composition of artifacts are of anthropological interest only if they can be related to problems of energetic requirements, resource use, technological change, production level, the organization of production and similar problems, and thus ultimately related to the role of technology as an aspect of culture.

Fifth, archaeologists cannot reasonably expect to master all the skills necessary to fully investigate metallurgical industries or other prehispanic technologies. Multidisciplinary studies, preferably in the context of the regional archaeological project mentioned above, are needed.

Colonial studies as a possible avenue to investigate Prehispanic technology

Research on the Early Colonial period also holds great potential for the study of the interaction between technology and culture in Mesoamerica. After the Spanish conquest, many traditions in Mesoamerica gradually atrophied and died out because there was no longer a place for them in the newly introduced socio-economic and political order. Other native traditions persisted minimally altered well into the post-Conquest period. They were practical ideas or patterns, or were so fundamental to the native mentality that they were not easily destroyed or changed. A great many other traditions, however, were functionally or formally transformed, enduring in new situations or under new guises. The old purposes they served changed, or their functions remained but their forms were modified or adapted to new conditions. In this way, traditional values or ideas endured through the media of European forms, and traditional forms shifted their context in order to convey what were essentially European ideas and values (e.g. Boone and Cummins 1998). The latter was the case of Mesoamerican metallurgical technology.

From the Middle Ages, the Western world had been experiencing constant growth in industry and trade. Throughout Europe, the production of metals steadily increased. Given the European climate that fostered an awareness of the importance of metals, it was inevitable that the search for ores would be carried to lands beyond the Atlantic, with an emphasis on scarce metals such as gold, silver and copper (Bargalló 1955; Mulholland 1981).

One of the more puzzling questions in the history of Colonial mining and metallurgy, however, is the reason for the delay in the emergence of large-scale metallurgical processes, particularly in the case of copper production in New Spain. The reason may well have been that the easy profits available to Spanish mine owners by leaving mining and refining to Indian workers made them unwilling to learn and invest in new techniques which required both new skills and substantial investment (Bargalló 1955). As a result, in the early decades of Colonial metal production in the Americas there was a broad and important continuation of pre-Conquest mining and smelting methods. In fact, the copper industry suffered a decline in productivity during the Early Colonial period. The low level of production apparently was associated with the continuing decline in indigenous population (Barret 1987; see also Chapter 4).

Because of its use in many critical sectors of the economy, copper was a significant resource in New Spain. In a society where iron goods were scarce, copper was indispensable. It was the most widely employed utilitarian metal in Spanish colonial times. Despite the Crown's need for copper for armament production, no move was made to take control of the mines from the Indians in the early post-Conquest period. The principal source of copper in New Spain throughout the Colonial period was the province of Michoacán, and the center of copper mining was the Real de Inguarán and other mines in the lower Balsas basin (Barret 1987).

This is significant since, as pointed out in Chapter 4, the Balsas region appears to have been the source of ore for the smelters around Itziparátzico. We are fortunate to have substantial documentary resources for colonial mining (e.g. Barret 1987; Warren 1968, 1989). The early colonial industry is understood largely through the documentary record; that is, through the accounts and reports of the Indian, mestizo and Spanish witnesses and the administrative records required by and still preserved in the Spanish and colonial administrative systems. Ethnohistoric sources, while valuable, may often carry inherent bias.

Unfortunately, almost no evidence comes from archaeology, because traditionally there has been little excavation of colonial contexts except as an overlay to a pre-Columbian site. A comprehensive investigation of an early colonial context would contribute new knowledge that will help us better understand the native industry. The focus may remain on the indigenous culture, but it could be directed toward understanding the recently changed post-Conquest world. In such a study, attention should be directed toward the indigenous response to the Spanish intrusion and the cultural adjustment and negotiation it required. In this manner, we may be able to better understand some of the cultural and socio-technological features that made these societies so resilient. We may also reveal essential traditions, technical frameworks, and ways of doing and organizing that might stand out more clearly after the Conquest than before.

Conclusions

To summarize, what generalizations can be made about the Tarascan copper industry on the basis of the combined archaeological, ethnohistorical and technological data presented in this volume?

First, contrary to preconceived notions of Mesoamerican metallurgy, the slag evidence at Itziparátzico revealed a sophisticated smelting method, possibly involving the use of furnaces properly designed and built to provide reducing conditions during the refinement of copper ores. The representations of metalworkers heating metal in crucibles by blowing through pipes, which are often found in ethnographic documents, may represent melting of metal ingots for final processing, rather than smelting of ores.

Second, not all the extractive processes were being carried out around the mining districts. Ores were transported to Itziparátzico and probably to a number of other similar Late Postclassic sites (e.g. Jilcalán El Viejo) exclusively for smelting. Final processing (alloying and fabrication, surface treatment, and finishing of metalwork) apparently was taking place at separate locations, likely, in the Tarascan political core.

This suggests that the copper industry was divided into minor sectors of production, including mining, smelting, and final processing, and probably a range of subdivisions within these.

Third and finally, state control over the production of copper in the Tarascan territory varied in degree and operated through different principles and mechanisms. Elite involvement in the final processing at the administrative centers, for instance, seems to have been more direct than it was in other sectors of the industry. Evidence of elite presence at Itziparátzico (supported by ethnographic data on land tenure) suggests, however, that some of the smelting activities may also have been under some form of state supervision. In the mining districts, on the other hand, activities were apparently organized around the seasonal labor of part-time metallurgical specialists. These were village residents engaged primarily in agriculture but who devoted some portion of the year to copper mining and smelting. The nature of demand and distribution may have changed in response to larger economic and political conditions, but production remained locally organized.

In spite of the above limitations and the related needs for future research, the results of this investigation provide valuable information about the metallurgical *chaîne opératoire* and the organization of technology and craft production in ancient Mesoamerica. The present study has defined and addressed a few selected problems of the technology and organization of Tarascan copper metallurgy. These problems are of both regional archaeological and wider theoretical interest.

It has become apparent from the incompleteness of the answers provided here, that further archaeological work is needed at Itziparátzico and in similar situations in the Tarascan region, as well as in the mining areas in the Balsas Basin. The threats of recent development projects to the archaeological evidence for technology and settlement lend some urgency to the need for additional research. A comprehensive regional investigation of settlement and metallurgical technology would allow further testing of some of the premises advanced in this study. Nevertheless, the present work has demonstrated the value of multi-approach studies. By juxtaposing multiple lines of evidence in an interdisciplinary manner, archaeologists can examine the distant and the local, the general and the particular, bringing ancient societies into better focus.

Bibliography

Acuña, René (ed.) 1985. *Relaciones Geográficas del Siglo XVI: Mexico*, vols. 6-8. Universidad Nacional Autónoma de Mexico, Mexico.

Adams, Richard N. 1975. *Energy and Structure: A Theory of Social Power*. University of Texas Press, Austin.

Aitchison, Leslie 1960. A History of Metals, 2 vols. Interscience Publishers, New York.

Alcalá de Jesús, María, Carlos A. Ortiz S., and Ma. del Carmen Gutiérrez C. 2001. Clasificación de los Suelos de la Meseta Tarasca, Michoacán. *Terra* 19 (3): 227-239.

Alpern, Stanley B. 2005. Did They or Didn't They Invent It? Iron in Sub-Saharan Africa. *History in Africa* 32: 41-94

Ames, Kenneth M. 1995. Chiefly Power and Household Production on the Northwest Coast. In *Foundations of Social Inequality*, edited T. Douglas Price and Gary M. Feinman, pp. 155-187. Plenum Press, New York.

Arnold, Dieter 1991. *Building in Egypt: Pharonic Stone Masonry*. Oxford University Press, New York.

Arnold, Jeanne E. and Ann Munns 1994. Independent or Attached Specialization: The Organization of Shell Bead Production in California. *Journal of Field Archaeology* 21: 473-489.

Arnold, Philip J., III 1996. Craft specialization and Social Change along the Southern Gulf Coast of Mexico. In *Craft Specialization and Social Evolution: In Memory of V. Gordon Childe*, edited by Bernard Wailes, pp. 201-210. University Museum Symposium Series, Volume 6. University Museum of Archaeology and Anthropology, University of Pennsylvania, Philadelphia.

Arsandaux, Henry and Paul Rivet 1921. Contribution à l'Etude de la Métallurgie Mexicaine. *Journal de la Société des Américanistes de Paris* 13: 261-280.

Bachmann, Hans-Gert 1982. *The Identification of Slags from Archaeological Sites*. Institute of Archaeology, Occasional Publication No. 6, London.

Bakewell, Peter J. 1984. *Miners of the Red Mountain: Indian Labor in Potosí, 1545-1650*. University of New Mexico Press, Albuquerque.

Bamforth, Douglas B. and Peter Bleed 1997. Technology, Flaked Stone Technology, and Risk. In *Rediscovering Darwin: Evolutionary Theory in Archaeological Explanation*, edited by C. M. Barton and G. A. Clark, pp. 109–139. Archaeological Papers of the American Anthropological Association 7, Washington.

Bargalló, Modesto 1955. *La Minería y la Metalurgia en la América Española durante la Época Colonial*. Fondo de Cultura Económica, Mexico.

Barnett, Homer G. 1953. *Innovation: The Basis of Cultural Change*. McGraw-Hill, New York.

Barrett, Elinore M. 1987. *The Mexican Colonial Copper Industry*. University of New Mexico Press, Albuquerque.

Bar-Yosef, O., B. Vandermeersch, B. Arensburg, A. Belfer-Cohen, P. Goldberg, H. Laville, L. Meignen, Y. Rak, J.D. Speth, E. Tchernov, A-M. Tillier, and S. Weiner 1992. The Excavations in Kebara Cave, Mt. Carmel. *Current Anthropology* 33 (5): 497-550.

Bennett, Wendell C. 1946. The Archaeology of Colombia. In *Handbook of South American Indians*, Vol. 2, edited by Julian H. Steward, pp. 823-864. Smithsonian Institution, Bureau of American Ethnology, Washington.

Berdan, Frances F. and Patricia R. Anawalt 1992. *The Codex Mendoza*. University of California Press, Berkeley.

Betancourt, Philip P. 2006. *Chrysokamino I: The Metallurgy Workshop and Its Territory*. Hesperia Supplement 36, American School of Classical Studies, Athens.

Binford, Lewis R. 1968. Some Comments on Historical versus Processual Archaeology. *Southwestern Journal of Anthropology* 24: 267-75.

Binford, Lewis R. 1989. Styles of Style. *Journal of Anthropological Archaeology* 8: 51-67.

Bird, Junius B. 1975. The 'Copper Man': A Prehistoric Miner and His Tools from Northern Chile. In *Pre-Columbian Metallurgy of South America*, edited by E. Benson, pp. 105-131. Dumbarton Oaks Research Library and Collections, Washington.

Blanton, Richard E. and Gary M. Feinman 1984. The Mesoamerican World System. *American Anthropologist* 86: 673-682.

Blanton, Richard E., Gary M. Feinman, Stephen Kowaleski, and Peter N. Peregrine 1996. A Dual-Processual Theory for the Evolution of Mesoamerican Civilization. *Current Anthropology* 37 (1): 1-14.

Blanton, Richard E., Stephen A. Kowalewski, Gary M. Feinman, and Laura M. Finsten 1993. *Ancient Mesoamerica: A Comparison of Change in Three Regions*, Second Edition. Cambridge University Press, Cambridge.

Bleed, Peter 1991. Operations Research and Archaeology. *American Antiquity* 56 (1): 19-35.

Bleed, Peter 1997. Content as Variability, Result as Selection: Toward a Behavioral Definition of Technology. *Archeological Papers of the American Anthropological Association* 7 (1): 95-104.

Bleed, Peter 2001. Trees or Chains, Links or Branches: Conceptual Alternatives for Consideration of Stone Tool Production and Other Sequential Activities. *Journal of Archaeological Method and Theory* 8 (1): 101-127.

Bober, Mandell M. 1955. *Karl Marx's Interpretation of History.* 2d ed., rev. Harvard University Press, Cambridge.

Boone, Elizabeth Hill, and Tom Cummins (eds) 1998. *Native Traditions in the Postconquest World.* Dumbarton Oaks Research Library and Collection, Washington.

Boone, James L. 1992. Competition, Conflict, and Development of Social Hierarchies. In *Evolutionary Ecology and Human Behavior,* edited by Eric A. Smith and Bruce Winterhalder, pp. 301-338. De Gruyter, New York.

Boserup, Ester 1965. *The Conditions of Agricultural Growth: The Economics of Agrarian Change Under Population Pressure.* Aldine Pub. Co., Chicago.

Bourdieu, Pierre 1977. *Outline of a Theory of Practice.* Cambridge University Press, Cambridge.

Boyd, Robert J. and Peter J. Richerson 1985. *Culture and the Evolutionary Process.* University of Chicago Press, Chicago.

Bradbury, R.E. 1957. *The Benin Kingdom and the Edo-Speaking Peoples of South-Western Nigeria.* International African Institute, London.

Bradbury, R.E. 1967. The Kingdom of Benin. In *West African Kingdoms in the Nineteenth Century,* edited Daryll C. Forde and Phyllis M. Kaberry, pp 1–35. International African Institute, Oxford University Press, London.

Bradley, Bruce A. 1982. Flaked Stone Technology and Typology. In *The Agate Basin Site: A Record of Paleoindian Occupation of the Northwest Great Plains,* edited by George C. Frison and Dennis J. Stanford, pp.181-208. Academic Press, NewYork.

Brand, Donald D. 1951. *Quiroga: A Mexican Municipio.* Smithsonian Institution, Institute of Social Anthropology No. 11. U.S. Govt. Print. Off., Washington.

Braun David P. 1990. Selection and Evolution in Nonhierarchical Organization. In *Evolution of Political Systems: Sociopolitics in Small-Scale Sedentary Societies,* edited by Steadman Upham, pp. 62–86. Cambridge University Press, Cambridge.

Bray, Warwick M. 1977. Maya Metalwork and its External Connections. In *Social Process in Maya Prehistory,* edited by Norman Hammond, pp. 365-403. Academic Press, London.

Broughton, Jack M., and James F. O'Connell 1999. On Evolutionary Ecology, Selectionist Archaeology and Behavioral Archaeology. (Response to Michael B. Schiffer, Vol. 61, p. 643, 1996). *American Antiquity* 64 (1): 153-165.

Brumfiel, Elizabeth M. 1991. Weaving and Cooking: Women's Production in Aztec Mexico. In *Engendering Archaeology: Women and Prehistory,* edited by Joan M. Gero and Margaret W. Conkey, pp. 224-254. Basil Blackwell, Oxford.

1992. Breaking and Entering the Ecosystem: Gender, Class, and Faction Steal the Show. *American Anthropologist* 94: 551-567.

Brumfiel, Elizabeth M. and Timothy K. Earle 1987. Specialization, Exchange and Complex Societies: An Introduction. In *Specialization, Exchange and Complex Societies,* edited by Elizabeth M. Brumfiel and Timothy K. Earle, pp. 1-9. Cambridge University Press, Cambridge.

Brumfiel, Elizabeth M. and John W. Fox (eds) 1994. *Factional Competition and Political Development in the New World.* Cambridge University Press, Cambridge.

Brush, Charles F. 1962. Pre-Columbian Alloy Objects from Guerrero, Mexico. *Science* 138 (3547): 1336-1338.

Budd, Paul and Barbara S. Ottaway 1995. Eneolithic Arsenical Copper: Chance or Choice? In *Ancient Mining and Metallurgy in Southeast Europe,* edited by Borislav Jovanović, pp. 95-102. Archaeological Institute/Museum of Mining and Metallurgy, Belgrade/Bor.

Burford, Alison 1972. *Craftsmen in Greek and Roman Society.* Cornell University Press, Ithaca.

Burger, Richard L. and Robert B. Gordon 1998. Early Central Andean Metalworking from Mina Perdida, Peru. *Science* 282 (5391): 1108-1111.

Cabrera C., Rubén 1986. El Desarrollo Cultural Prehispánico en el Bajo Río Balsas. In *Arqueología y Etnohistoria del Estado de Guerrero,* pp. 117-151. Instituto Nacional de Antropología e Historia, Mexico.

Capoche, Luis, 1546. or 7-1613. 1959. *Relaciones Histórico-Literarias de la América Meridional: Relación General de la Villa Imperial de Potosí.* Ediciones Atlas, Madrid.

Cárdenas G., Efraín 1986. Registro de Sitios del Centro de Michoacán. Informe inédito en el Archivo Técnico de la Dirección de Registro Arqueológico, INAH, Mexico.

Cárdenas G., Efraín and Eugenia Fernández-Villanueva M. 2004. Metalurgia en Santa Clara del Cobre: ¿Una Tradición Prehispánica? In *Ritmo del Fuego: El Arte y los Artesanos de Santa Clara del Cobre, Michoacán, Mexico,* edited by Michele Feder-Nadoff, pp. 92-101. Fundación Cuentos, Chicago.

Carneiro, Robert L. 1974. A Reappraisal of the Roles of Technology and Organization in the Origin of Civilization. *American Antiquity* 39 (2): 179-186.

Carranza A., Mario, Javier López A., Alba E. Pérez R., and Carlos F. Yáñez M. (eds) 1995. *Geological-Mining Monograph of the State of Michoacán.* Consejo de Recursos Minerales, Centro Minero, Pachuca.

Carrillo-Cázares, Alberto 1996. *Partidos y Padrones del Obispado de Michoacán 1680-1685.* El Colegio de Michoacán, Zamora

Chadwick, Robert 1971. Archaeological Synthesis of Michoacan and Adjacent Areas. In *Handbook of Middle American Indians, Vol. 11, Archaeology of Northern Mesoamerica Part 2,* edited by Gordon F. Ekholm and Ignacio Bernal, General Editor Robert Wauchope, pp. 657-693. University of Texas Press, Austin.

Charnov, Eric L. 1976. Optimal Foraging: The Marginal Value Theorum. *Theoretical Population Biology* 9: 129-136.

Chayanov, Alexander V. 1986. *The Theory of Peasant Economy*. University of Wisconsin Press, Madison.

Chazan, Michael 2004. Locating Gesture: Leroi-Gourhan among the Cyborgs. Gestures, Rituals and Memory: A Multidisciplinary Approach to Patterned Human Movement across Time, Virtual Symposium: http://www.semioticon.com

Childe, V. Gordon 1930. *The Bronze Age*. Cambridge University Press, Cambridge.

Childe, V. Gordon 1942. *What Happened in History*. Pelican Books, Hammondsworth.

Childe, V. Gordon 1944. Archaeological Ages as Technological Stages. *Journal of the Royal Anthropological Institute of Great Britain and Ireland* 44 (74): 7-24.

Childe, V. Gordon 1946. The Social Implications of the Three 'Ages' in Archaeological Classification. *The Modern Quarterly* N.S. 1: 18-33.

Childe, V. Gordon 1950. The Urban Revolution. *Town Planning Review* 23: 3-17.

Childe, V. Gordon 1951a *Man Makes Himself*. New American Library, New York.

Childe, V. Gordon 1951b *Social Evolution*. World Publishing Company, New York.

Childe, V. Gordon 1954. Early Forms of Society. In *A History of Technology*, Vol. I, edited by Charles J. Singer, E. J. Holmyard, A. R. Hall, and Trevor I. Williams, pp. 38-54. Oxford University Press, Oxford.

Childe, V. Gordon 1958. *The Prehistory of European Society*. Penguin Books, Harmondsworth.

Childs S. Terry and David J. Killick 1993. Indigenous African Metallurgy: Nature and Culture. *Annual Review of Anthropology* 22: 317-337.

Cieza de León, Pedro de, 1518-1554. 1985. *Crónica del Perú. Segunda Parte*. Pontificia Universidad Católica del Perú, Fondo Editorial, Academia Nacional de la Historia, Lima.

Clarke, David L.1968. *Analytical Archaeology*. Methuen, London.

Clark, John E. 1986. From Mountains to Molehills: A Critical Review of Teotihuacan's Obsidian Industry. In *Economic Aspects of Prehispanic Highland Mexico*, edited by Barry L. Isaac, pp. 23-74. Research in Economic Anthropology, Supplement 2. JAI Press, Greenwich.

Clark, John E. and Michael Blake 1994. The Power of Prestige: Competitive Generosity and the Emergence of Rank Societies in Lowland Mesoamerica. In *Factional Competition and Political Development in the New World*, edited by E. M. Brumfiel and J. W. Fox, pp. 17-30. Cambridge University Press, Cambridge.

Cline, Walter B. 1937. *Mining and Metallurgy in Negro Africa*. George Banta Publishing Company, Menasha.

Coghlan, Herbert H. 1942. Some Fresh Aspects on the Prehistoric Metallurgy of Copper. *Antiquaries Journal* 22: 22-40.

Coghlan, Herbert H. 1951. *Notes on the Prehistoric Metallurgy of Copper and Bronze in the Old World*. Pitt Rivers Museum Occasional Papers on Technology 4. University Press, Oxford.

Coghlan, Herbert H. 1956. *Notes on Prehistoric and Early Iron in the Old World*. Pitt Rivers Museum Occasional Papers on Technology 8. University Press, Oxford.

Coghlan, Herbert H. 1972. Some Reflections on the Prehistoric Working of Copper and Bronze. *Archaeologia Austriaca* 52: 93-104.

Collins, Michael B. 1975. Lithic Technology as a Means of Processual Inference. In *Lithic Technology: Making and Using Stone Tools*, edited by Earl Swanson, pp. 15-34. Mouton Publishers, The Hague.

Connah, Graham 1975. *The Archaeology of Benin: Excavations and Other Researches in and Around Benin City, Nigeria*. Clarendon Press, Oxford.

Costin, Cathy L. 1991. Craft Specialization: Issues in Defining, Documenting, and Explaining the Organization of Production. In *Archaeological Method and Theory*, edited by Michael B. Schiffer, pp. 1-56. University of Arizona Press, Tucson.

Costin, Cathy L. 1996. Craft Production and Mobilization Strategies in the Inka Empire. In *Craft Specialization and Social Evolution: In Memory of V. Gordon Childe*, edited by Bernard Wailes, pp. 211-225. The University of Pennsylvania Museum, Philadelphia.

Costin, Cathy L. 2001. Craft Production Systems. In *Archaeology at the Millennium: A Sourcebook*, edited by Gary M. Feinman and T. Douglas Price, pp. 273-327. Kluwer Academic/Plenum Publishers, New York.

Costin, Cathy L. and Rita P. Wright (eds) 1998. *Craft and Social Identity*. Archaeological Papers of the American Anthropological Association No. 8, Washington.

Cottrell, Alan H., Sir 1995. *An Introduction to Metallurgy*, Second Edition. The Institute of Materials, London.

Cowgill, George L. 1997. State and Society at Teotihuacan, Mexico. *Annual Review of Anthropology* 26: 129-161.

Crabtree, Don E. 1966. A Stoneworker's Approach to Analyzing and Replicating the Lindenmeier Folsom. *Tebiwa* 9 (1): 3-39.

Crabtree, Don E. 1968. Mesoamerican Polyhedral Cores and Prismatic Blades. *American Antiquity* 33 (4): 446-478.

Craddock, Paul T. 1980. *Scientific Studies in Early Mining and Extractive Metallurgy*. British Museum Occasional Paper 20, London.

Craddock, Paul T. 1991. Mining and Smelting in Antiquity. In *Science and the Past*, edited by Sheridan Bowman, 57-73. University of Toronto Press, Toronto and Buffalo.

Craddock, Paul T. 2001. From Hearth to Furnace: Evidences for the Earliest Metal Smelting Technologies in the Eastern Mediterranean. *Paléorient* 26 (2): 151-165.

Craine, Eugene R. and Reginald C. Reindorp 1970. *The Chronicles of Michoacán.* University of Oklahoma Press, Norman.

Cruz A., Rafael and Alejandro Pastrana C. 1994. Sierra Las Navajas, Hidalgo: Nuevas Investigaciones sobre la Explotación Pre-Colonial de Obsidiana. In *Simposium sobre Arqueología en el Estado de Hidalgo: Trabajos recientes (1989),* editado por Enrique Fernández D., pp. 31-45). Instituto Nacional de Antropología e Historia, Mexico.

D'Altroy, Terrance N. and Timothy K. Earle 1985. Staple Finance, Wealth Finance, and Storage in the Inka Political Economy. *Current Anthropology* 26 (2): 187-206.

Daniel, Glyn E. 1964. *The Idea of Prehistory.* Penguin Books, Harmondsworth.

Daniel, Glyn E. and Christopher Chippindale 1989. *The Pastmasters : Eleven Modern Pioneers of Archaeology : V. Gordon Childe, Stuart Piggott, Charles Phillips, Christopher Hawkes, Seton Lloyd, Robert J. Braidwood, Gordon R. Willey, C.J. Becker, Sigfried J. De Laet, J. Desmond Clark, D.J. Mulvaney.* Thames and Hudson, New York.

Davies, Sarah J., Sarah E. Metcalfe, Fernando Bernal-Brooks, Arturo Chacón-Torres, John G. Farmer, A. B. MacKenzie, and Anthony J. Newton 2005. Lake Sediments Record Sensitivity of Two Hydrologically Closed Upland Lakes in Mexico to Human Impact. *AMBIO: A Journal of the Human Environment* 34 (6): 470–475.

Davies, Sarah J., Sarah E. Metcalfe, Angus B. MacKenzie, Anthony J. Newton, Georgina H. Endfield and John G. Farmer 2004. Environmental Changes in the Zirahuén Basin, Michoacán, Mexico, During the Last 1000. Years. *Journal of Paleolimnology* 31(1): 77-98.

DeMarrais, Elizabeth, Luis Jaime Castillo, and Timothy Earle 1996. Ideology, Materialization, and Power Strategies. *Current Anthropology* 37 (1): 15-30.

Desch, Cecil H. 1936. Analyses of Metal. In *Excavations at Thermi in Lesbos,* by Winifred Lamb. The University press, Cambridge.

Diamond, Jared 1999. *Guns, Germs, and Steel: The Fates of Human Societies.* W.W. Norton & Company, New York.

Downing, Theodore E. and McGuire Gibson (eds) 1974. *Irrigation's Impact on Society.* Anthropological Papers of the University of Arizona No. 25. University of Arizona Press, Tucson.

Drennan, Robert D. 1984. Long-Distance Movement of Goods in the Mesoamerican Formative and Classic. *American Antiquity* 49 (1): 27-43.

Duncan, Lynne C. 1999. Roman Deep-Vein Mining. University of North Carolina: http://earthsci.org/mineral/mindep/ancient_mine/deep-vein_mining.htm

Dunnell, Robert C. 1978. Style and Function: A Fundamental Dichotomy. *American Antiquity* 43 (2):192-202.

Dunnell, Robert C. 1980. Evolutionary Theory and Archaeology. In *Advances in Archaeological Method and Theory,* Vol. 3, edited by Michael B. Schiffer, pp. 35-99. Academic Press, New York.

Dunnell, Robert C. 1982. Science, Social Science, and Common Sense: The Agonizing Dilemma of Modern Archaeology. *Journal of Anthropological Research* 38: 1-25.

Dupree, A. Hunter 1969. The Role of Technology in Society and the Need for Historical Perspective. *Technology and Culture* 10: 528-34.

Earle, Timothy K. 1997. *How Chiefs Come to Power: The Political Economy in Prehistory.* Stanford University Press, Stanford.

Edmonds, Mark 1990. Description, Understanding and the Chaîne Opératoire. *Archaeological Review from Cambridge* 9 (1): 55-70.

Edwards, Clinton R 1969. Possibilities of Pre-Columbian Maritime Contacts among New World Civilizations. In *Precolumbian Contact within Nuclear America,* Mesoamerican Studies vol. 4, edited by J. Charles Kelley and Carroll L. Riley, pp. 3-10. Southern Illinois University Press, Carbondale.

Effenberg, Günter and Rainer Schmid-Fetzer 2000. *Critical Evaluation of Ternary Phase Diagram Data.* MSI, Stuttgart.

Endfield, Georgina H. and Sarah L. O'Hara 1999. Degradation, Drought, and Dissent: An Environmental History of Colonial Michoacán, West Central Mexico. *Annals of the Association of American Geographers* 89 (3): 402-41.

Ensor, Bradley E., Marisa O. Ensor, and Gregory W. De Vries 2003. Hohokam Political Ecology and Vulnerability: Comments on Waters and Ravesloot. *American Antiquity* 68 (1): 169-181.

Epstein, Stephen M. 1993. Cultural Choice and Technological Consequences: Constraint of Innovation in the Late Prehistoric Copper Smelting Industry of Cerro Huaringa, Peru. Ph.D. Dissertation, University of Pennsylvania, Philadelphia.

Etchevers B., Jorge D. 1985. *Un Cuarto de Siglo de Investigación en Suelos Volcánicos de Mexico.* Serie Cuadernos de Edafología 1. Centro de Edafología, Colegio de Postgraduados, Chapingo.

Feinman, Gary M. 1999. Rethinking Our Assumptions: Economic Specialization at the Household Scale in Ancient Ejutla, Oaxaca, Mexico. In *Pottery and People: Dynamic Interactions,* edited by James M. Skibo and Gary Feinman, pp. 81-98. University of Utah Press, Salt Lake City.

Feinman, Gary M. and Jill Neitzel 1984. Too Many Types: An Overview of Sedentary Pre-State Societies in the Americas. In *Advances in Archaeological Method and Theory,* Vol. 7, edited by Michael B. Schiffer, pp. 39-102. Academic Press, New York.

Feinman, Gary M. and Linda M. Nicholas 1991. New Perspectives on Prehispanic Highland Mesoamerica: A Macroregional Approach. *Comparative Civilizations Review* 24:13-33.

Feinman, Gary M. and Linda M. Nicholas 2000. Household Craft Specialization and Shell Ornament Manufacture in Ejutla, Mexico. In *Exploring the Past: Readings in Archaeology*, edited by James M. Bayman and Miriam T. Stark, pp. 303-314. Carolina Academic Press, Durham.

Ferrari Luca, Victor H. Garduno, Giorgio Pasquarè and Alessandro Tibaldi 1994. Volcanic and Tectonic Evolution of Central Mexico: Oligocene to Present. *Geofísica Internacional* 33 (1): 91-105.

Fitzhugh, J. Benjamin 2001. Risk and Invention in Human Technological Evolution. *Journal of Anthropological Archaeology* 20: 125–167.

Flannery, Kent V. 1972. The Cultural Evolution of Civilizations. *Annual Review of Ecology and Systematics* 3: 399-426.

Flenniken, J. Jeffrey 1978. Reevaluation of the Lindenmeier Folsom: A Replication Experiment in Lithic Technology. *American Antiquity* 43 (3): 473-479.

Forbes, Robert J. 1950. *Metallurgy in Antiquity: A Notebook for Archaeologists and Technologists.* E. J. Brill, Leiden.

Forbes, Robert J. 1955-1972. *Studies in Ancient Technology*, 12 Volumes. E. J. Brill, Leiden.

Ford, Anabel 1996. Critical Resource Control and the Rise of the Classic Period Maya. In *The Managed Mosaic: Ancient Maya Agriculture and Resource Use*, edited by Scott L. Fedick, pp. 297-303. University of Utah Press, Salt Lake City.

Giddens, Anthony 1979. *Central Problems in Social Theory: Action, Structure and Contradictions in Social Analysis.* University of California Press, Berkeley.

Giddens, Anthony 1984. *The Constitution of Society: Outline of the Theory of Structuration.* University of California Press, Berkeley.

Gilead, Isaac and Steve Rosen 1992. New Archaeo-Metallurgical Evidence for the Beginnings of Metallurgy in the Southern Levant: Excavations at Tell Abu Matar, Beersheba (Israel) 1990/1. *Institute for Archaeo-Metallurgical Studies (IAMS)* 18: 11-14.

Golden, Jonathan, Thomas E. Levy, and Andreas Hauptmann 2001. Recent Discoveries Concerning Chalcolithic Metallurgy at Shiqmim, Israel. *Journal of Archaeological Science* 28: 951-963.

Gorenstein, Shirley and Helen. P. Pollard 1983. *The Tarascan Civilization: A Late Prehispanic Cultural System.* Publications in Anthropology 28, Nashville.

Grace, Roger 1996. The 'Chaîne Opératoire' Approach to Lithic Analysis. Stone Age Reference Collection, Hypertexual Publications: http://www.hf.uio.no/iakh/forskning/sarc/iakh/lithic/sarc.html

Green, Sally 1981. *Prehistorian: A biography of V. Gordon Childe.* Moonraker, Bradford-on-Avon.

Greenberg, James B. and Thomas K. Park 1994. Political Ecology. *Journal of Political Ecology* 1: 1-12.

Grinberg, Dora M.K. de 1989. Tecnologías Metalúrgicas Tarascas. *Ciencia y Desarrollo* 15 (89): 37-52.

Grinberg, Dora M.K. de 1990. *Los Señores del Metal. Minería y Metalurgia en Mesoamérica.* Dirección General de Publicaciones del CNCA/Pangea, Mexico.

Grinberg, Dora M.K. de 1995. El Legajo 1204. del Archivo General de Indias, el Lienzo de Jucutacato y las Minas Prehispánicas de Cobre del Ario, Michoacán. In *Arqueología del Norte y Occidente de Mexico*, edited by Barbo Dahlgren y Ma. Dolores Soto de Arechavaleta, pp. 211-265. UNAM, Instituto de Investigaciones Antropológicas, Mexico.

Grinberg, Dora M.K. de 1996. Técnicas Minero-Metalúrgicas en Mesoamérica. In *Mesoamérica y los Andes*, edited by Mayán Cervantes, pp. 427-471. Centro de Investigaciones y Estudios Superiores de Antropología Social, Mexico.

Grinberg, Dora M.K. de 2004. ¿Qué Sabían de Fundición los Antiguos Habitantes de Mesoamérica? *Ingenierías* VII (22): 64-70.

Grinberg, Dora M.K. de and Francisca Franco V. 1987. Estudio de Cuatro Cascabeles de Falso Alambre Provenientes de las Excavaciones del Tren Subterráneo de la Ciudad de Mexico. *Antropología y Técnica* 2: 143-151.

Guevara F., Fernando 2004. El Escenario Ecológico de la Región de Santa Clara del Cobre. In *Ritmo del Fuego: El Arte y los Artesanos de Santa Clara del Cobre, Michoacán, Mexico*, edited by Michele Feder-Nadoff, pp. 66-89. Fundación Cuentos, Chicago.

Habashi, Fathi 1986. *Principles of Extractive Metallurgy*, Vol. 3, Pyrometallurgy. Gordon and Breach, New York.

Habashi, Fathi 2005. Fire and the Art of Metals: A Short History of Pyrometallurgy. *Mineral Processing and Extractive Metallurgy, IMM Transactions section C* 114 (3): 165-171.

Harris, David R. (ed.) 1994. *The Archaeology of V. Gordon Childe: Contemporary Perspectives.* Proceedings of the V. Gordon Childe Centennial Conference held at the Institute of Archaeology, UCL, 1992. University of Chicago Press, Chicago.

Harris, Marvin 1968. *The Rise of Anthropological Theory.* Thomas Crowell, New York.

Harwell, Robert 1962. A Revolution in the Chinese Iron and Coal Industries During the Northern Sung, 960-1126. A.D. *Journal of Asian Studies* 21: 153-162.

Harwell, Robert 1966. Markets, Technology, and the Structure of Enterprise in the Development of the Eleventh Century Chinese Iron and Steel Industry. *Journal of Economic History* 26: 29-58.

Hasenaka, Toshi and Ian S.E. Carmichael 1985. The Cinder Cones of Michoacan-Guanajuato: Their Age, Volume and Distribution, and Magma Discharge Rate. *Journal of Volcanology and Geothermal Research* 25: 105-124.

Hassig, Ross 1985. *Trade, Tribute, and Transportation: The Sixteenth-Century Political Economy of the Valley of Mexico.* University of Oklahoma Press, Norman.

Hauptmann, Andreas 2003. Developments in Copper Metallurgy During the Fourth and Third Millennia BC at Feinan, Jordan. In *Mining and Metal Production through the Ages*, edited by Paul T. Craddock and Janet Lang, pp. 90-100. The British Museum, London.

Hauptmann, Andreas, Thilo Rehren and Sigrid Schmitt-Strecker 2003. Early Bronze Age Copper Metallurgy at Shahr-i Sokhta (Iran), Reconsidered. In *Man and Mining - Mensch und Bergbau*, edited by Th. Stöllner, G. Körlin, G. Steffens and J. Cierny, pp. 197-213. Der Anschnitt, Beiheft 16, Deutsches Bergbau-Museum, Bochum.

Hay, Conran A. 1978. Kaminaljuyu Obsidian: Lithic Analysis and the Economic Organization of a Prehistoric Mayan Chiefdom. Ph.D. Dissertation, Department of Anthropology, The Pennsylvania State University, University Park.

Hayashida, Frances M. 1995. State Pottery Production in the Inka Provinces. Ph.D. Dissertation, Department of Anthropology, University of Michigan, Ann Arbor.

Hayden, Brian 1994. Competition, Labor, and Complex Hunter-Gatherers. In *Key Issues in Hunter-Gatherer Research*, edited by Ernest S. Burch and Linda J. Ellanna, pp. 223-239. Berg Press, Oxford.

Hayden, Brian 1995. Pathways to Power: Principles for Creating Socioeconomic Inequalities. In *Foundations of Social Inequality*, edited by T. Douglas Price and Gary M. Feinman, pp. 15-85. Plenum Press, New York.

Healy, John F. 1977. *Mining and Metallurgy in the Greek and Roman World*. Thames and Hudson, London.

Hegmon, Michelle, Winston Hurst, and James R. Allison 1995. Production for Local Consumption and Exchange: Comparisons of Early Red and White Ware Ceramics in the San Juan Region. In *Ceramic Production in the American Southwest*, edited by Barbara J. Mills and Patricia L. Crown, pp. 30-62. University of Arizona Press, Tucson.

Helms, Mary W. 1993. *Craft and the Kingly Ideal*. University of Texas Press, Austin.

Henderson, Hope 2003. The Organization of Staple Crop Production at K'axob, Belize. *Latin American Antiquity* 14 (4): 469-96.

Henderson, Julian 2000. *The Science and Archaeology of Materials: An Investigation of Inorganic Materials*. Routledge, London and New York.

Hendrichs, Pedro 1940. Datos sobre la Técnica Minera Prehispánica. *Mexico Antiguo* 5: 148-160, 179-194, 311-238.

Hendrichs, Pedro 1943- Por *Tierras Ignotas: Viajes y Observaciones en la Región del Río de las Balsas*. 2 1944. vols. Editorial Cultura, Mexico.

Hers, Marie-Areti 1990. Los Objetos de Cobre en la Cultura Chalchihuites. In *Homenaje a Federico Sescosse: Un Hombre, Un Destino y un Lugar*, pp. 45-60. Gobierno del estado de Zacatecas, Zacatecas.

Hirth, Kenneth G. 1992. Interregional Exchange as Elite Behavior: An Evolutionary Perspective. In *Mesoamerican Elites: An Archaeological Assessment*, edited by Diane Z. Chase and Arlen F. Chase, pp. 18-29. University of Oklahoma Press, Norman.

Hirth, Kenneth G. 1996. Political Economy and Archaeology: Perspectives on Exchange and Production. *Journal of Archaeological Research* 4 (3): 203-39.

Hirth, Kenneth G. (ed.) 2003. *Mesoamerican Lithic Technology: Experimentation and Interpretation*. The University of Utah Press, Salt Lake City.

Hirth, Kenneth G. and Bradford Andrews (eds) 2002. *Pathways to Prismatic Blades: A Study in Mesoamerican Obsidian Core-Blade Technology*. University of California, Los Angeles.

Hobson, Robert L 1962. *The Wares of the Ming Dynasty*. C.E. Tuttle, Rutland.

Hodges, Henry 1970. *Technology in the Ancient World*. Knopf, New York.

Hoover, Herbert C. and L. Henry Hoover 1950. *Georgius Agricola: De Re Metallica*. Translated from the first Latin edition of 1556. Dover Publications Inc., New York.

Horcasitas de Barros, Maria Luisa 1981. *Una Artesanía con Raíces Prehispánicas en Santa Clara del Cobre*. Colección Etnología, INAH, Mexico.

Horne, Lee 1982. Fuel for the Metal Worker. *Expedition* 25 (1): 8-13.

Hosler, Dorothy 1988a Ancient West Mexican Metallurgy: South Central American Origins and West Mexican Transformations. *American Anthropologist* 90 (4): 832-855.

Hosler, Dorothy 1988b Ancient West Mexican Metallurgy: A Technological Chronology. *Journal of Field Archaeology* 15 (2): 191-217.

Hosler, Dorothy 1988c The Metallurgy of Ancient West Mexico. In *The Beginning of the Use of Metals and Alloys*, edited by Robert Maddin, pp. 328–343. MIT Press, Cambridge.

Hosler, Dorothy 1994. *The Sounds and Colors of Power*. MIT Press, Cambridge.

Hosler, Dorothy 1995. Sound, Color and Meaning in the Metallurgy of Ancient West Mexico. *World Archaeology* 27: 100-115.

Hosler, Dorothy 1996. Technical Choices, Social Categories and Meaning among the Andean Potters of Las Animas. *Journal of Material Culture* 1(l): 63-92.

Hosler, Dorothy 2002. Excavations at the Copper Smelting Site of El Manchon, Guerrero, Mexico. Reports Submitted to FAMSI: http://www.famsi.org/reports/01058/index.html

Hosler, Dorothy, Heather Lechtman, and Olaf Holm 1990. *Axe-Monies and Their Relatives*. Studies in Pre-Columbian Art and Archaeology. 30, Dumbarton Oaks, Washington.

Hosler, Dorothy and Andrew Macfarlane 1996. Copper Sources, Metal Production, and Metals Trade in Late Postclassic Mesoamerica. *Science* 273 (5283): 1819-1824.

Hunt, Robert C. and Hunt, Eva 1976. Canal Irrigation and Local Social Organization. *Current Anthropology* 17(3): 389-411.

INEGI (Instituto Nacional de Estadística, Geografía e Informática) 2000. XII Censo General de Población y Vivienda 2000. Sistemas Nacionales Estadístico y de Información Geográfica, Mexico: *http://www.inegi.gob.mx/inegi/default.asp*

Inomata, Takeshi 2001. The Power and Ideology of Artistic Creation: Elite Craft Specialists in Classic Maya Society. *Current Anthropology* 42 (3): 321-49.

Ixer, Robert A. 1999. The Role of Ore Geology and Ores in the Archaeological Provenancing of Metals. In *Metals in Antiquity*, edited by Suzanne M.M. Young, A. Mark Pollard, Paul Budd and Robert A. Ixer, pp. 43-52. BAR International Series 792, Archaeopress, Oxford.

Janusek, John W. 1999. Craft and Local Power: Embedded Specialization in Tiwanaku Cities. *Latin American Antiquity* 10 (2): 107-131.

Jones, George T., Robert D. Leonard and Alysia L. Abbott 1995. The Structure of Selectionist Explanations in Archaeology. In *Evolutionary Archaeology: Methodological Issues*, edited by Patrice A. Teltser, pp. 13-32. University of Arizona Press, Tucson.

Joyce, Arthur A., Laura Arnaud Bustamante, and Marc N. Levine 2001. Commoner Power: A Case Study from the Classic-Period Collapse on the Oaxaca Coast. *Journal of Archaeological Theory and Method* 8: 343-85.

Joyce, Rosemary A. 2000. High culture, Mesoamerican civilization, and the Classic Maya tradition. In *Order, Legitimacy, and Wealth in Ancient States*, edited by Janet Richards and Mary Van Buren, pp. 64–76. Cambridge University Press, Cambridge.

Juleff, Gillian 1996. An Ancient Wind-powered Iron Smelting Technology in Sri Lanka. *Nature* 379 (3): 60-63.

Juleff, Gillian 1998a *Early Iron and Steel in Sri Lanka: A Study of the Samanalawewa Area*. AVA Materialien 54, Verlag Philipp von Zabern, Mainz.

Juleff, Gillian 1998b Ancient Iron and Steel Production at Samanalawewa. *Sabaragamuwa University Journal* 1 (1): 3-9.

Keiller, Alexander, Stuart Piggott and F.S. Wallis 1941. First Report of the Sub-Committee of the South-Western Group of Museums and Art Galleries on the Petrological Identification of Stone Axes. *Proceedings of the Prehistoric Society* 7: 50-72.

Killick, David J. 1991. The Relevance of Recent African Iron Smelting Practice to Reconstructions of Prehistoric Smelting Technology. In *Recent Trends in Archaeometallurgical Research*, edited by Petar D. Glumac, pp. 47-54. MASCA Research Papers in Science and Archaeology 8, part 1. University of Pennsylvania, Philadelphia.

Killick, David J. 1996. On claims for 'Advanced' Ironworking Technology in Precolonial Africa. In *The Culture and Technology of African Iron Production*, edited by Peter R. Schmidt, pp. 247-266. University Press of Florida, Gainesville.

Krebs, John R. and Nicholas B. Davies 1993. *An Introduction to Behavioural Ecology*. Blackwell Scientific Publications, Oxford.

Kuhn, Steven L. 2004. Evolutionary Perspectives on Technology and Technological Change. *World Archaeology* 36 (4): 561-570.

Langenscheidt, Adolphus 1970. *Minería Prehispánica en la Sierra de Querétaro*. Secretaría del Patrimonio Nacional, Mexico.

Langenscheidt, Adolphus 1985. Bosquejo de la Minería Prehispánica en Mexico. *Quipu* 2 (1): 37-57.

Langenscheidt, Adolphus 1988. *Historia Mínima de La Minería en la Sierra Gorda*. Rolston-Bain, Mexico.

Lechtman, Heather 1973. The Gilding of Metals in Pre-Columbian Peru. In *Application of Science in Examination of Works of Art*, edited by William J. Young, pp. 38-52. Museum of Fine Arts, Boston.

Lechtman, Heather 1977. Style in Technology: Some Early Thoughts. In *Material Culture: Styles, Organization, and Dynamics of Technology*, edited by Heather N. Lechtman and Robert S. Merrill, pp. 3-20. West Publishing, St. Paul.

Lechtman, Heather 1979. Issues in Andean Metallurgy. In *Pre-Columbian metallurgy of South America*, edited by Elizabeth P. Benson, pp. 1-40. Dumbarton Oaks, Washington.

Lechtman, Heather 1980. The Central Andes: Metallurgy without Iron. In *The Coming of the Age of Iron*, edited by Theodore A. Wertime and James D. Muhly, pp. 267-334. Yale University Press, New Haven.

Lechtman, Heather 1985. The Manufacture of Copper-Arsenic Alloys in Prehistory. *Historical Metallurgy* 19 (1): 41-142.

Lechtman, Heather 1993. Technologies of Power: the Andean Case. In *Configurations of Power in Complex Society: Holistic Anthropology in Theory and Practice*, edited by John S. Henderson and Patricia J. Netherly, pp. 244-280. Cornell University Press, Ithaca.

Lechtman, Heather 2003. Tiwanaku Period (Middle Horizon) Bronze Metallurgy in the Lake Titicaca Basin: A Preliminary Assessment. In *Tiwanaku and Its Hinterland* Vol. 2: Urban and Rural Archaeology, edited by Alan L. Kolata, pp. 404–434. Smithsonian Institution Press, Washington.

Lechtman, Heather and Ana María Soldi (eds) 1981. *Runakunap Kawsayninkupaq Rurasqankunaqa: La Tecnología en el Mundo Andino*. Universidad Nacional Autónoma de Mexico, Instituto de Investigaciones Antropológicas, Mexico.

Lee, Richard B. and Irven DeVore (eds) 1968. *Man the Hunter*. Aldine Pub. Co., New York.

Lemonnier, Pierre 1989. Bark Capes, Arrowheads and the Concorde: On Social Representations of Technology. In *The Meanings of Things: Material*

Culture and Symbolic Expression, edited by Ian Hodder, pp. 156-171. Unwin Hyman, London.

Lemonnier, Pierre 1992. *Elements for an Anthropology of Technology*. Anthropological Papers of the Museum of Anthropology No. 88, University of Michigan, Ann Arbor.

Leonard, Robert D. and George T. Jones 1987. Elements of an Inclusive Evolutionary Model for Archaeology. *Journal of Anthropological Archaeology* 6:199–219.

Leonard, Robert D. and Heidi E. Reed 1993. Population Aggregation in the Prehistoric American Southwest: A Selectionist Model. *American Antiquity* 58: 648-661.

León-Portilla, Miguel 1978. *La Minería en Mexico: Estudios Sobre su Desarrollo Histórico*. Universidad Nacional Autónoma de Mexico, Mexico.

Leopold, A. Starker 1950. Vegetation Zones of Mexico. *Ecology* 31:507-18.

Leroi-Gourhan, André G. 1964. *Le Geste at la Parole*. Albin Michelle, Paris.

Lohse, Jon C.
2004. Blue Creek Regional Political Ecology Project 2003. Season Summary – Project Overview. In *2003. Season Summaries of the Blue Creek Regional Political Ecology Project, Upper Northwestern Belize*, edited by Jon C. Lohse, pp. 1-14. Maya Research Program, Ft. Worth, and The University of Texas, Austin.

Loney, Heather L. 2000. Society and Technological Control: A Critical Review of Models of Technological Change in Ceramic Studies. *American Antiquity* 65(4): 646-668.

Long, Stanley V. 1964. *Cire Perdue* Copper Casting in Pre-Columbian Mexico: An Experimental Approach. *American Antiquity* 30 (2): 189-192.

Longacre, William A. 1968. Some Aspects of Prehistoric Society in East Central Arizona. In *New Perspectives in Archeology*, edited by Sally R. Binford and Lewis R. Binford, pp.89–102. Aldine Pub. Co., Chicago.

Lucas, Alfred and John R. Harris 1962. *Ancient Egyptian Materials and Industries*. 4th ed., revised by John R. Harris. Edward Arnold and Company, London.

Lyman, R. Lee and Michael J. O'Brien 1997. The Concept of Evolution in Early Twentieth Century Americanist Archeology. In *Rediscovering Darwin: Evolutionary Theory and Archaeological Explanation*, edited by C. Michael Barton and Geoffrey A. Clark, pp. 21-48. Archeological Papers of the American Anthropological Association 7, Arlington.

Macías G., Angelina 1990. *Huandacareo: Lugar de Juicios, Tribunal*. Serie Arqueología, Colección Científica Núm. 222, Instituto Nacional de Antropología e Historia, Mexico.

Maldonado, Blanca E. 2001. Copper Production In Santa Clara Del Cobre, Michoacan. Paper presented at the SAA meetings, New Orleans. Unpublished Manuscript.

Maldonado, Blanca E. 2002. Modern Metallurgy, Prehispanic Roots: Coppersmithing in Mexico. Paper presented at the SAA Meetings, Denver. Unpublished Manuscript.

Maldonado, Blanca E. 2005. Análisis Tecnológico de la Metalurgia Prehispánica de Michoacán. In *Ethnoarqueología, El Contexto Dinámico de la Cultura Material A Través del Tiempo*, edited by Eduardo Williams, pp. 215-235. El Colegio de Michoacán, A.C., Zamora.

Maldonado, Blanca E., Thilo Rehren and Paul R. Howell 2005. Archaeological Copper Smelting at Itziparátzico, Michoacan, Mexico. *Materials Issues in Art and Archaeology VII*, edited by Pamela B. Vandiver, Jennifer L. Mass, Alison Murray, pp. 231-240. MRS Proceedings Volume 852, Warrendale.

Maldonado C., Rubén 1980. *Ofrendas Asociadas a Entierros del Infiernillo en el Balsas: Estudio y Experimentación con Tres Métodos de Taxonomía Numérica*. Instituto Nacional de Antropología e Historia, Mexico.

Manzanilla, Linda (ed.) 1987. *Studies in the Neolithic and Urban Revolutions: The V. Gordon Childe Colloquium, Mexico, 1986*. BAR International Series, Oxford.

Martínez de Lejarza, Juan J. 1974. *Análisis Estadístico de la Provincia de Michoacán en 1822*. Colección 'Estudios Michoacanos' IV, Fimax, Morelia.

Martínez S., Manuel (ed.) 1903. *Relación de las Ceremonias y Ritos y Población y Gobernación de los Indios de la Provincia de Mechuacan hecha al Ilustrísimo Señor Don Antonio de Mendoza, Virrey y Gobernador de esta Nueva España por Su Majestad*. Tip. de A. Aragon, Morelia.

Marx, Karl 1904. *A Contribution to the Critique of Political Economy*. Translated from the 2d German ed. by Nahum I. Stone. Charles H. Kerr & company, Chicago.

Marx, Karl 1955. *Capital*. Edited by Friedrich Engels. Translated from the 3d German ed. by Samuel Moore and Edward Aveling. Revised, with additional translation from the 4th German ed., by Marie Sachey and Herbert Lamm. Encyclopaedia Britannica, Chicago.

Masson, Marilyn A. and David A. Freidel (eds) 2002. *Ancient Maya Political Economies*. Alta Mira Press, Lanham.

Matson, Frederick R. (ed.) 1965a *Ceramics and Man*. Viking Fund Publications in Anthropology 41. Aldine Pub. Co., Chicago.

Matson, Frederick R. 1965b Ceramic Ecology: An approach to the Study of Early Cultures in the Near East. In *Ceramics and Man*, edited by Frederick R. Matson, pp. 202-217. Viking Fund Publications in Anthropology 41. Aldine Pub. Co., Chicago.

McNairn, Barbara 1980. *The Method and Theory of V. Gordon Childe*. Edinburgh University Press, Edinburgh.

Meighan, Clement W. 1969. Cultural Similarities Between Western Mexico and Andean Regions. In *Precolumbian Contact within Nuclear America*, Mesoamerican Studies vol. 4, edited by J. Charles Kelley and Carroll L. Riley, pp. 11-25. Southern Illinois University Press, Carbondale.

Merkel, John F. and Beno Rothenberg 1999. The Earliest Steps to Copper Metallurgy in the Western Arabah. *The Beginnings of Metallurgy: Proceedings of the International Conference 'The Beginnings of Metallurgy', Bochum 1995*, edited by Andreas Hauptmann, Ernst Pernicka, Thilo Rehren, and Ünsal Yalcin, 149-165. Dt. Bergbau-Museum, Bochum.

Merrill, Robert S. 1968. The Study of Technology. *International Encyclopedia of the Social Sciences* 15: 577-589.

Meyerdirks, U., Thilo Rehren, and A. Harvey 2004. Reconstructing the Early Medieval Copper Smelting at Ross Island. In *Ross Island. Mining, Metal and Society in Early Ireland*, ed. William O'Brien, pp.651-664. National University of Ireland, Galway.

Miller, Duncan 2003. Indigenous Copper Mining and Smelting in Pre-Colonial Southern Africa. In *Mining and Metal Production through the Ages*, edited by Paul T. Craddock and Janet Lang, pp. 101-110. The British Museum, London.

Morán Z, Dante J. 1984. *Geología de la República Mexicana*. SPP-INEGI, Mexico.

Moreno G., Heriberto 1986. *Diario del Viaje a la Nueva España, Francisco de Ajofrín*. Secretería de Educación Pública, Mexico.

Moreno H., Auxilio, F. Contreras C., J. A. Cámara S., and J. L. Simón G. 2003. Metallurgical Control and Social Power: The Bronze Age Communities of High Guadalquivir (Spain). In *Archaeometallurgy in Europe*. Proceedings of the International Conference (Milan, September 24-26, 2003). Associazione Italiana di Metallurgia, Vol. 1, pp. 625-634, Milan.

Mountjoy, Joseph B. 1969. On the Origin of West Mexican Metallurgy. In *Precolumbian Contact within Nuclear America*, Mesoamerican Studies vol. 4, edited by J. Charles Kelley and Carroll L. Riley, pp. 26-42. Southern Illinois University Press, Carbondale.

Mountjoy, Joseph B. and Luis Torres M. 1985. The Production and Use of Prehispanic Metal Artifacts in the Central Coastal Area of Jalisco, Mexico. In *The Archaeology of West and Northwest Mesoamerica*, edited by Michael S. Foster and Phil C. Weigand, pp. 133-152. Westview Press, Boulder.

Muan, Arnulf 1957. Phase Equilibria at Liquidus Temperatures in the System Iron oxide- Al_2O_3- SiO_2 in Air Atmosphere. *Journal of the American Ceramic Society* 40 (4): 121-133.

Mulholland, James A. 1981. *A History of Metals in Colonial America*. University of Alabama Press, Birmingham.

Múnera B., L. Carlos and Saburo Sugiyama 1998. Ritual Ceramics at a Workshop in the Ciudadela, Teotihuacán: Catalog. Reports Submitted to FAMSI: http://www.famsi.org/reports/97050/section02a.htm

Needham, N. Joseph 1958. *The Development of Iron and Steel Technology in China*. Second Dickinson Memorial Lecture to the Newcomen Society. The Newcomen Society, London.

Netting, Robert M. 1977. *Cultural Ecology*. Cummings Publishing Company, Reading.

Netting, Robert M. 1993. *Smallholders, Householders: Farm Families and the Ecology of Intensive, Sustainable Agriculture*. Stanford University Press, Stanford.

Noble, Joseph V. 1965. *The Techniques of Painted Attic Pottery*. Watson-Guptill Publications, New York.

O'Brian, Michael J. and Thomas D. Holland 1990. Variation, Selection, and the Archeological Record. *Archaeological Method and Theory* 2: 31-79.

O'Brian, Michael J. and Thomas D. Holland 1992. The Role of Adaptation in Archaeological Explanation. *American Antiquity* 57 (1): 36-59.

O'Brian, Michael J. and Thomas D. Holland 1995. Behavioral Archaeology and the Extended Phenotype. In *Expanding Archaeology*, edited by James M. Skibo, William H. Walker, and Axel E. Nielsen, pp. 143-161. University of Utah Press, Salt Lake City.

Oddy, William A. (ed.) 1977. *Aspects of Early Metallurgy*. British Museum Occasional Paper No 17, London.

Olin, Jacqueline S. and Alan D. Franklin (eds) 1982. *Archaeological Ceramics*. Smithsonian Institution Press, Washington.

Ostroumov, Mikhail and Pedro Corona-Chávez 2000. Yacimientos Minerales en Michoacán: Aspectos Geológicos y Metalogenéticos. *Revista Ciencia Nicolaita* 23: 7-22.

Ostroumov, Mikhail, Pedro Corona Chávez, Jorge Díaz de León, Alfredo Victoria Morales and Juan Carlos Cruz Ocampo. 2002. Taxonomía y Clasificación Cristaloquímica Moderna de los Minerales. Recursos Electrónicos de la Universidad Michoacana: http://smm.iim.umich.mx/catalogo.htm

Oswalt, Wendell H. 1982. Material Culture in Anthropology. In *Culture and Ecology: Eclectic Perspectives*, edited by John G. Kennedy and Robert B. Edgerton, pp. 56–64. Special Publication No. 15, American Anthropological Association, Washington.

Ottaway, Barbara S. 1994. *Prähistorische Archäometallurgie*. Leidorf, Espelkamp.

Ottaway, Barbara S. 2001. Innovation, Production and Specialization in Early Prehistoric Copper Metallurgy. *European Journal of Archaeology* 4 (1): 87-112.

Paredes M., Carlos S. 1984. El Tributo Indígena en la Región del Lago de Pátzcuaro. In *Michoacán en el Siglo XVI*, edited by Carlos S. Paredes M., pp. 21-104. Colección Estudios Michoacanos VII, FIMAX Publicistas, Morelia.

Paredes M., Carlos S. 2004. Fragmentos de la Historia de Santa Clara del Cobre y del Imperio Español en la Época Colonial. In *Ritmo del Fuego: El Arte y los Artesanos de Santa Clara del Cobre, Michoacán, Mexico*, edited by Michele Feder-Nadoff, pp. 154-161. Fundación Cuentos, Chicago.

Parsons, Jeffrey R, Keith W. Kintigh and Susan A Gregg 1983. *Archaeological Settlement Pattern Data from the Chalco, Xochimilco, Ixtapalapa, Texcoco, and Zumpango Regions, Mexico*. Museum of Anthropology Technical Reports No. 14, University of Michigan, Ann Arbor.

Patterson, Thomas C. 2005. Craft Specialization, the Reorganization of Production Relations and State Formation. *Journal of Social Archaeology* 5 (3): 307-337.

Pauketat, Timothy R. 2001. Practice and History in Archaeology: An Emerging Paradigm. *Anthropological Theory* 1:73-98.

Paynter, Robert and Randall H. McGuire 1991. The Archaeology of Inequality: Material Culture, Domination, and Resistance. In *The Archaeology of Inequality*, edited by Randall H. McGuire and Robert Paynter, pp. 1-27. Basil Blackwell, Cambridge.

Pearson, Harry W. 1957. The Economy Has No Surplus. In *Trade and Market in the Early Empires: Economies in History and Theory*, edited by Karl Polanyi, Conrad M. Arensberg and Harry W. Pearson, pp. 320-341. The Free Press, Glencoe.

Peele, Robert Jr. 1893. A Primitive Smelting Furnace. *School of Mines Quarterly* 15: 8-10.

Pendergast, David M. 1962. Metal Artifacts in Prehispanic Mesoamerica. *American Antiquity* 27 (4): 520-545.

Peregrine, Peter 1991. Some Political Aspects of Craft Specialization. *World Archaeology* 23:1-11.

Pérez-Cálix, Emmanuel 1996. *Flora y Vegetación de la Cuenca del Lago de Zirahuén, Michoacán, Mexico*. Flora del Bajío y de Regiones Adyacentes 13, Fascículo Complementario. Centro Regional del Bajío del Instituto de Ecología, A.C., Pátzcuaro.

Petersen, Georg 1970. *Minería y Metalurgia en el Antiguo Perú*. Arqueológicas 12, Museo Nacional de Antropología y Arqueología, Lima.

Pollard, Helen P. 1982. Ecological Variation and Economic Exchange in the Tarascan State. *American Ethnologist* 9 (2): 250-268.

Pollard, Helen P. 1987. The Political Economy of Prehispanic Tarascan Metallurgy. *American Antiquity* 52 (4):741-752.

Pollard, Helen P. 1993. *Tariacuri's Legacy: The Prehispanic Tarascan State*. University of Oklahoma Press, Norman.

Pollard, Helen P. 1995. Estudio del Surgimiento del Estado Tarasco: Investigaciones Recientes. In *Arqueología del Occidente y Norte de Mexico*, edited by Eduardo Williams and Phil C. Weigand, pp. 29-64. El Colegio de Michoacán, Zamora.

Pollard, Helen P. 1997. Recent Research in West Mexican Archaeology. *Journal of Archaeological Research* 5 (4): 345-384.

Pollard, Helen P. 2000. Tarascan External Relationships. In *Greater Mesoamerica: The Archaeology of West and Northwest Mexico*, edited by Michael S. Foster and Shirley Gorenstein, pp. 71-80. University of Utah Press, Salt Lake City.

Pollard, Helen P. 2005. Desarrollo del Estado Tarasco: Proyecto Erongarícuaro. Informe Técnico Parcial de la Temporada 1 (Campo), 2001. y la Temporada 2 (Laboratorio), 2002-2004.

Pollard, Helen P. and Laura Cahue 1999. Mortuary Patterns of Regional Elites in the Lake Pátzcuaro Basin of Western Mexico. *Latin American Antiquity* 10 (3): 259-280.

Pollard, A. Mark and Carl Heron 1996. *Archaeological Chemistry*. Royal Society of Chemistry, Cambridge.

Ponomarenko, Alyson L. 2004. The Pachuca Obsidian Source, Hidalgo, Mexico: A Geoarchaeological Perspective. *Geoarchaeology* 19 (1): 71–91.

Price, Barbara J. 1982. Cultural Materialism: A Theoretical Review. *American Antiquity* 47(4):709-741.

Quilter, Jeffrey 1998. Metallic Reflections. *Science* 282 (5391): 1058-1059.

Raber, Paul A. 1984. The Organization and Development of Early Copper Metallurgy in the Polis Region, Western Cyprus. Ph.D. Dissertation, Department of Anthropology, The Pennsylvania State University, University Park.

Rathje, William L. 1971. The Origin and Development of Lowland Classic Maya Civilization. *American Antiquity* 36 (3): 275-85.

Rattray, Evelyn C. 1987. La Producción y la Distribución de Obsidiana en el Período Coyotlatelco en Teotihuacan. In *Teotihuacan: Nuevos Datos, Nuevos Síntesis, Nuevos Problemas*, edited by Emily McClung de Tapia and Evelyn C. Rattray, pp. 451-463. Instituto de Investigaciones Antropológicas, Arqueología. Serie Antropológica 72, UNAM, Mexico.

Redman, Charles L. 1978. *The Rise of Civilization*. W.H. Freeman and Company, San Francisco.

Reents-Budet, Dorie 1998. Elite Maya Pottery and Artisans as Social Indicators. In *Craft and Social Identity*, edited by Cathy Costin and Rita Wright, pp. 71-89. Archaeological Papers of the American Anthropological Association No. 8, Washington.

Rehder, John E. 994. Blowpipes versus Bellows in Ancient Metallurgy. *Journal of Field Archaeology* 21: 345-350.

Renfrew, A. Colin 1972. *The Emergence of Civilisation: The Cyclades and the Aegean in the Third Millennium B.C.* Methuen, London.

Renfrew, A. Colin 1979. *Before Civilization: The Radiocarbon Revolution and Prehistoric Europe*. Cambridge University Press, New York.

Renfrew, A. Colin, J.E. Dixon and J.R. Cann 1966. Obsidian and Early Culture Contact in the Near East. *Proceedings of the Prehistoric Society* 32: 30-72.

Renfrew, A. Colin and Malcolm Wagstaff (eds) 1982. *An Island Polity: The Archaeology of Exploitation in Melos*. Cambridge University Press, Cambridge.

Renfrew, A. Colin and Malcolm Wagstaff (eds) 1968. Further Analysis of Near East Obsidian. *Proceedings of the Prehistoric Society* 34: 319-331.

Rice, Prudence M.1991. Specialization, Standardization, and Diversity: A Retrospective. In *The Ceramic Legacy of Anna O. Shepard*, edited by Ronald L. Bishop and Frederick W. Lange, pp. 257-279. University of Colorado Press, Boulder.

Robbins, Paul 2004. *Political Ecology*. Blackwell Publishing, Oxford.

Roskamp, Hans 2001. Historia, Mito y Legitimación: El Lienzo de Jicalán. In *La Tierra Caliente de Michoacán*, edited by Eduardo Zárate Hernández, pp.119-151. El Colegio de Michoacán/Gobierno del Estado de Michoacán, Zamora.

Roskamp, Hans 2003. Los Códices de Cutzio y Huetamo: Encomienda y Tributo en la Tierra Caliente de Michoacán, siglo XVI. Colegio de Michoacán/Colegio Mexiquense, Zamora.

Roskamp, Hans 2004. Los Caciques Indígenas de Xiuhquilan y la Defensa del las Minas en el Siglo XVI: El Lienzo de Jicalán. In *Ritmo del Fuego: El Arte y los Artesanos de Santa Clara del Cobre, Michoacán, Mexico*, edited by Michele Feder-Nadoff, pp. 186-197. Fundación Cuentos, Chicago.

Roskamp, Hans, Mario Retiz, Anyul Cuellar, and Efraín Cárdenas 2003. Pre-Hispanic and Colonial Metallurgy in Jicalán, Michoacán, Mexico: An Archaeological Survey. Reports Submitted to FAMSI: http://www.famsi.org/reports/02011/index.html

Rostoker, William, Vincent C. Pigott, and James R. Dvorak 1989. Direct Reduction to Copper Metal by Oxide-Sulfide Mineral Interaction. *Archeomaterials* 3 (1): 69-87.

Roth, H. Ling 1968. *Great Benin: Its Customs, Art and Horrors*. Originally 1903. Routledge & Kegan Paul, London.

Rothenberg, Beno 1985. Copper Smelting Furnaces in the Arabah, Israel: The Archeological Evidence. In *Furnaces and Smelting-Technology in Antiquity*, edited by Paul T. Craddock and Michael J. Hughes, pp. 123-150. British Museum Occasional Paper No. 48. British Museum Publications, London.

Rothenberg, Beno and John F. Merkel 1995. Late Neolithic Copper Smelting in the Arabah. *Institute for Archaeo-Metallurgical Studies (IAMS)* 19: 1-7.

Rothenberg, Beno and C. Timothy Shaw 1990. The Discovery of a Copper Mine and Smelter from the End of the Early Bronze Age (EBIV) in the Timna Valley. *Institute for Archaeo-Metallurgical Studies (IAMS)* 15/16: 1-8.

Roux, Valentine 1989. *The Potter's Wheel: Craft Specialization and Technical Competence*. Oxford and IBH Publishing, New Delhi.

Roux, Valentine, and Pierre Matarasso 1999. Crafts and the Evolution of Complex Societies: New Methodologies for Modelling the Organization of Production, a Harappan Example. In *The Social Dynamics of Technology, Practice, Politics and World Views*, edited by Marcia-Anne Dobres and Christopher R. Hoffman, pp. 46-70. Smithsonian Institution Press, Washington and London.

Rubín de la Borbolla, Daniel F. 1944. Orfebrería Tarasca. *Cuadernos Americanos* 3: 125-138.

Ryder, Alan F. C. 1969. *Benin and the Europeans, 1485-1897*. Humanities Press, New York.

Sackett, James R. 1982. Approaches to Style in Lithic Archaeology. *Journal of Anthropological Archaeology* 1: 59-112.

Sackett, James R. 1986a Style, Function and Assemblage Variability: A Reply to Binford. *American Antiquity* 51 (3): 628-634.

Sackett, James R. 1986b Isochrestism and Style: A Clarification. *Journal of Anthropological Archaeology* 5: 266-277.

Sackett, James R. 1990. Style and Ethnicity in Archaeology: The Case for Isochrestism. In *The Uses of Style in Archaeology*, edited by M. W. Conkey and C. Hastorf, pp. 32-43. Cambridge University Press, Cambridge.

Sahagún, Fray Bernardino de 1969- *Florentine Codex: General History of the Things of New Spain, 1590*. 12 books.

Sahagún, Fray Bernardino de 1982. Translated and edited by Arthur J.O. Anderson and Charles E. Dibble. School of American Research and the University of Utah Press, Santa Fe and Salt Lake City.

Sahlins, Marshall 1972. *Stone Age Economics*. Aldine, Atherton, Inc., Chicago.

Sanders, William T. 1977. Environmental Heterogeneity and the Evolution of Lowland Maya Civilization. In *The Origins of Maya Civilization*, edited by Richard E. W. Adams, pp. 287-97. University of New Mexico Press, Albuquerque.

Sanders, William T. and Barbara J. Price 1968. *Mesoamerica: The Evolution of a Civilization*. Random House, New York.

Sanders, William, Jeffrey Parsons, and Robert Santley 1979. *The Basin of Mexico: Ecological Process in the Evolution of a Civilization*. Academic Press, New York.

Sanders, William T. and David L. Webster 1978. Unilinealism, Multilinealism, and the Evolution of Complex Societies. In *Social Archaeology: Beyond Subsistence and Dating*, edited by Charles L. Redman, pp. 249-301. Academic Press, New York

Santley, Robert S. 1994. Specialized Commodity Production in and around Matacapan: Testing the Goodness of Fit of the Regal-Ritual and Administrative Models. In *Archaeological Views from the Countryside: Village Communities in Early Complex Societies*, edited by Glen M. Schwartz and Steven E. Falconer, pp. 91-108. Smithsonian Institution Press, Washington.

Santley, Robert S. 2004. Prehistoric Salt Production at El Salado, Southern Veracruz, Mexico. *Latin American Antiquity* 15 (2): 199-221.

Scarre, Chris J. 1999. Archaeological Theory in France and Britain. *Antiquity* 73 (279): 155-161.

Schiffer, Michael B. 1976. *Behavioral Archaeology.* Academic Press, New York.

Schiffer, Michael B. 1996. Some Relationships between Behavioral and Evolutionary Archaeologies. *American Antiquity* 61 (4): 643-662.

Schiffer, Michael B. and James M. Skibo 1987. Theory and Experiment in the Study of Technological Change. *Current Anthropology* 28 (5): 595-622.

Schiffer, Michael B. and James M. Skibo 1997. The Explanation of Artifact Variability. *American Antiquity* 62 (1): 27-50.

Schiffer, Michael B., James M. Skibo, Janet Griffits, Kacy Hollenback, and William A. Longacre 2001. Behavioral Archaeology and the Study of Technology. *American Antiquity* 66 (4): 729-738.

Schlanger, Nathan 1994. Mindful Technology: Unleashing the Chaine Operatoire for an Archaeology of Mind. In *The Ancient Mind: Elements of Cognitive Archaeology,* edited by A. Colin Renfrew and Ezra B. Zubrow, pp.143-151. Cambridge University Press, Cambridge.

Schmidt, Peter R. 1997. *Iron Technology in East Africa: Symbolism, Science, and Archaeology.* Indiana University Press, Bloomington.

Schmidt, Peter R. and Donald H. Avery 1978. Complex Iron Smelting and Prehistoric Culture in Tanzania. *Science* 201 (4361): 1085-1089.

Schortman, Edward M. and Patricia A. Urban 2004. Modeling the Roles of Craft Production in Ancient Political Economies. *Journal of Archaeological Research* 12 (2): 185-226.

Scott, David A. 1992. Spanish Colonial and Indigenous Platinum: A Historical Account of Technology and Fabrication. In *Conservation of the Iberian and Latin American Cultural Heritage: Preprints of the Contributions to the Madrid Congress,* edited by Henry W.M. Hodges, John S.Mills and Perry Smith, pp. 148-153. International Institute for Conservation of Historic and Artistic Works, London.

Scott, David A. 2002. The Gold Metallurgy of La Tolita: Some Achievements and Questions. Paper Presented at the Primer Symposium Internacional sobre Tecnología del Oro Antiguo: Europa y América, Madrid: *http://www.ih.csic.es/arqueometalurgia/sitoaprog.htm*

Scott, David A. and Warwick Bray 1980. Ancient Platinum Technology in South America. *Platinum Metals Review* 24: 144-157.

Scott, David A. and Warwick Bray 1994. Pre-Hispanic Platinum Alloys: Their Composition and Use in Ecuador and Colombia. In *Archaeometry of Pre-Columbian Sites and Artifacts: Proceedings of the 28th International Symposium on Archaeometry,* edited by David.A. Scott and Pieter Myers, pp. 285-322. Getty Conservation Institute, Los Angeles.

Sellet, Frédéric 1994. *Chaîne Opératoire*: The Concept and its Application. *Lithic Technology* 18: 106-112.

Service, Elman R. 1971. *Cultural Evolutionism: Theory in Practice.* Holt, Rinehart and Winston, New York.

Shackleton, William G. 1986. *Economic and Applied Geology: An introduction.* Croom Helm, London

Shalev, Sariel and Peter J. Northover 1987. Chalcolithic Metal and Metalworking from Shiqmim. In *Shiqmim I: Studies Concerning Chalcolithic Societies in the Northern Negev Desert, Israel (1982-1984),* edited by Thomas E. Levy, pp. 357-371. BAR International Series 356, Oxford.

Shanks, Michael, and Christopher Tilley 1992. *Re-Constructing Archaeology: Theory and Practice.* Second edition. Routledge Press, New York.

Sheets, Payson 2000. Provisioning the Household: the Vertical Economy, Village Economy, and Household Economy in the Southeastern Maya Periphery. *Ancient Mesoamerica* 11(2): 217-30.

Shepard, Anna O. 1940. *Rio Grande Glaze Paint Ware: A Study Illustrating the Place of Ceramic Technological Analysis in Archaeological Research.* Carnegie Institution of Washington, Publication 528, Contributions to American Anthropology and History 39, Washington.

1956. *Ceramics for the Archaeologist.* Carnegie Institution of Washington, Washington.

Shimada, Izumi, Stephen M. Epstein and Alan K. Craig 1982. Batán Grande: A Prehistoric Metallurgical Center in Peru. *Science* 216 (4549): 952-959.

Shimada, Izumi, Stephen M. Epstein and Alan K. Craig 1983. The Metallurgical Process in Ancient North Peru. *Archaeology* 35 (5): 38-45.

Shimada, Izumi and John F. Merkel 1991. Copper-Alloy Metallurgy in Ancient Peru. *Scientific American* 265 (1): 80-86.

Shimada, Izumi, Jo Anne Griffin and Adon Gordus 2000. The Technology, Iconography and Social Significance of Metals: A Multidimensional Analysis of Middle Sican Objects. In *Precolumbian Gold: Technology, Style and Iconography,* edited by C. McEwan, pp. 28-61. Fitzroy Dearborn Publishers, Chicago.

Simmons, Scott E. 2005. Maya Metallurgy and the Political Economy of Late Postclassic-Spanish Colonial Lamanai, Belize. Poster presented at the SAA Meetings, Salt Lake City.

Singer, Charles J, E. J. Holmyard, A. R. Hall, and Trevor I. Williams (eds) 1954- *A History of Technology,* 5 Volumes. Clarendon Press, Oxford.

1959. Sinopoli, Carla M. 1988. The Organization of Craft Production at Vijayanagara, South India. *American Anthropologist* 90 (3): 580-597.

1959. Sinopoli, Carla M. 2003. *The Political Economy of Craft Production: Crafting Empire in South India, c. 1350-1650.* Cambridge University Press, Cambridge.

Skibo, James M. and Michael B. Schiffer 2001. Understanding Artifact Variability and Change:

A Behavioral Framework. In *Anthropological Perspectives on Technology*, edited by M. B. Schiffer, pp. 139-149. Amerind Foundation, Dragoon.

Smith, Michael E. 2004. The Archaeology of Ancient State Economies. *Annual Review of Anthropology* 33: 73-102.

Solanes C., María del Carmen and Enrique Vela 2000. *Atlas del Mexico Prehispánico*. Revista Arqueología Mexicana, Núm. Especial 3. Editorial Raíces, Mexico.

Solis O., Felipe R. 1999. Arte Funerario en el Occidente de Mexico durante la Época Prehispánica. Correo del Maestro 42: http://www.correodelmaestro.com/multimedia/multimedia.htm

Spence, Michael W. 1981. Obsidian Production and the State in Teotihuacan. *American Antiquity* 46 (4): 769-88.

Spence, Michael W. 1982. The Social Context of Production and Exchange. In *Contexts for Prehistoric Exchange*, edited by Jonathon E. Ericson and Timothy K. Earle, p.p. 173-197. Academic Press, New York.

Spence, Michael W. 1984. Craft Production and Polity in Early Teotihuacan. In *Trade and Exchange in Early Mesoamerica*, edited by Kenneth G. Hirth, pp. 87-114. University of New Mexico Press, Albuquerque.

Spence, Michael W. 1987. The Scale and Structure of Obsidian Production in Teotihuacan. In *Teotihuacan: Nuevos Datos, Nuevos Síntesis, Nuevos Problemas*, edited by Emily McClung de Tapia and Evelyn C. Rattray, pp.429-50. Instituto de Investigaciones Antropológicas, Arqueología. Serie Antropológica 72, UNAM, Mexico.

Spence, Michael W. 1996. Commodity or Gift: Teotihuacan Obsidian in the Maya Region. *Latin American Antiquity* 7 (3): 21–39.

Spier, Robert F. G. 1970. *From the Hand of Man: Primitive and Preindustrial Technologies*. Houghton Mifflin, Boston.

Stark, Miriam T. (ed.) 1998. *The Archaeology of Social Boundaries*. Smithsonian Institution Press, Washington.

Stein, Gil J. 1996. Producers, Patrons, and Prestige: Craft Specialist and Emergent Elites in Mesopotamia from 5500-3100. B.C. In *Craft Specialization and Social Evolution: In Memory of V. Gordon Childe*, edited by Bernard Wailes, pp. 25-38. The University of Pennsylvania Museum, Philadelphia.

Stein, Gil J. 2001. Understanding Ancient State Societies in the Old World. In *Archaeology at the Millennium*, edited by Gary M. Feinman and T. Douglas Price, pp. 353–379. Kluwer Academic, New York.

Stein, Gil J. 2002. From Passive Periphery to Active Agents: Emerging Perspectives in the Archaeology of interregional Interaction. *American Anthropologist* 104 (3): 903-16.

Steinberg, Arthur 1975. Technology and Culture: Technological Styles in the Bronzes of Shang China, Phrygia and Urnfield Central Europe. In *Material Culture: Styles, Organization, and Dynamics of Technology*, edited by Heather N. Lechtman and Robert S. Merrill, pp. 53-86. West Publishing, St. Paul.

Steward, Julian H. 1949. Cultural Causality and Law: A Trial Formulation of the Development of Early Civilization. *American Anthropologist* 51: 1-27.

Steward, Julian H. 1951. Levels of Sociocultural Integration: An Operational Concept. *Southwestern Journal of Anthropology* 7: 374-90.

Steward, Julian H. 1955. *Theory of Culture Change: The Methodology of Multilinear Evolution*. University of Illinois Press, Urbana.

Sugiyama, Saburo 1998. Archaeology and Iconography of Theater-type Censers: Official Military Emblems from the Ciudadela? *Teotihuacan Notes: Internet Journal for Teotihuacan Archaeology and Iconography* I-2: http://archaeology.asu.edu/teo/notes/SS/noteI_2SS.htm

Teltser, Patrice A. 1995. Culture History, Evolutionary Theory, and Frequency Seriation. In *Evolutionary Archaeology: Methodological Issues*, edited by Patrice A. Teltser, pp. 151-168. University of Arizona Press, Tucson.

Tite, Michael S., Y. Maniatis, N.D. Meeks, M. Bimson, M.J. Hughes, and Leppard, S. C. 1982. Technological Studies of Ancient Ceramics from the Near East, Aegean and Southeast Europe. In *Early Pyrotechnology: The Evolution of the First Fire-Using Industries*, edited by Theodore A. Wertime and Steven F. Wertime, pp. 61-71. Smithsonian Institution Press, Washington.

Torrence, Robin 1989. Re-tooling: Towards a Behavioral Theory of Stone Tools. In *Time, Energy and Stone Tools*, edited by Robin Torrence, pp. 57-66. Cambridge University Press, Cambridge.

Torres M, Luis and Francisca Franco V. 1996. *La Metalurgia Tarasca: Producción y Uso de los Metales en Mesoamérica*. Editorial Instituto Nacional para la Cultura y las Artes, Mexico.

Tottle, Charles R. 1984. *An Encyclopædia of Metallurgy and Materials*. Macdonald and Evans, Plymouth.

Trigger, Bruce G. 1980. *Gordon Childe, Revolutions in Archaeology*. Columbia University Press, New York.

Trigger, Bruce G. 2003. *Understanding Early Civilizations: A Comparative Study*. Cambridge University Press, New York.

Tylecote, Ronald F. 1962. *Metallurgy in Archaeology: A Prehistory of Metallurgy in the British Isles*.

E. Arnold, London. 1976. *A History of Metallurgy*. Metals Society, London.

E. Arnold, London. 1986. *The Prehistory of Metallurgy in the British Isles*. The Institute of Metals, London.

Tylecote, Ronald F., Ghaznavi H.A. and P.J. Boydell 1977. Partitioning of Trace Elements Between the Ores, Fluxes, Slags and Metal During the Smelting of Copper. *Journal of Archaeological Science* 4 (4): 305-333.

Van Buren, Mary, W. E. Brooks, P. Calla, L. Cayo, S. deFrance, J. Eighmy, C. Eylar, S. Fidel, D. Goldstein, A. M. Presta, Th. Rehren, H. Stinchfield, B. Mills, and C. R. Cohen 2006. Proyecto Arqueológico Porco-Potosí: Inka and Spanish Colonial Silver Mining in the Bolivian Andes. Colorado State University, Knowledge to Go Places: *http://lamar.colostate.edu/~mvanbure/index.htm*

Van der Leeuw, Sander E. 1977. Towards a Study of the Economics of Pottery Making. In *Ex Horreo*, edited by B. L. van Beek, R. W. Brandt and W. Groenman-van Wateringe, pp. 68-76. Cingula IV. Albert Egges van Giffen Instituut voor Parae-en Protohistorie, Universiteit van Amsterdam, Amsterdam.

Van der Leeuw, Sander E. 1984. Dust to Dust: A Transformational View of the Ceramic Cycle. In *The Many Dimensions of Pottery*, edited by Sander E. van der Leeuw and Alison Pritchard, pp. 709-773. Albert Egges van Giffen Instituut voor Prae-en Prtohistoire, CINGULA VII, Universitat van Amsterdam, Amsterdam.

Van der Leeuw, Sander E. 1993. Giving the Potter a Choice: Conceptual Aspects of Pottery Techniques. In *Technological Choices: Transformation in Material Culture from the Neolithic to Modern High Tech*, edited by Pierre Lemonnier, pp. 238-288. Routledge, London.

Van der Leeuw, Sander E. 1994. Cognitive Aspects of Technique. In *The Ancient Mind: Elements of Cognitive Archaeology*, edited by A. Colin Renfrew and Ezra B. Zubrow, pp. 135-142. Cambridge University Press, Cambridge.

Veldhuijzen, Harald A. 2003. 'Slag_Fun', A New Tool for Archaeometallurgy: Development of an Analytical (P) ED-XRF Method for Iron-Rich Materials. *Papers from the Institute of Archaeology* 14: 102-118.

Wailes, Bernard (ed.) 1996. *Craft Specialization and Social Evolution: In Memory of V. Gordon Childe*. University Museum Symposium Series, Volume 6. University Museum of Archaeology and Anthropology, University of Pennsylvania, Philadelphia.

Warren, J. Benedict 1968. Minas de Cobre de Michoacán, 1533. *Anales del Museo Michoacano* 6: 35-52.

Warren, J. Benedict 1985. *The Conquest of Michoacán: The Spanish Domination of the Tarascan Kingdom in Western Mexico, 1521-1530*. University of Oklahoma Press, Norman.

Warren, J. Benedict 1989. Información del Licenciado Vasco de Quiroga sobre el Cobre de Michoacán, 1533. *Anales del Museo Michoacano* 1: 30-52.

Warren, J. Benedict 1991. *El Diccionario Grande de la Lengua de Michoacán*, II Tomos. Fimax, Morelia.

Webster, David L., Susan Toby Evans and William T. Sanders 1993. *Out of the Past: An Introduction to Archaeology*. Mayfield Publishing Co., Mountain View.

Weigand, Phil C. 1982. Introduction to *Mining and Mining Techniques in Ancient Mesoamerica*, edited by Phil C. Weigand and Gretchen Gwynne, pp. 1-6. Anthropology, 6: (1 and 2). Ukw, Stoney Brook.

Weiner, Annette B. 1976. *Women of Value, Men of Renown: New Perspectives in Trobriand Exchange*. University of Texas Press, Austin.

Welsch, Robert L. and John E. Terrell 1998. Material Culture, Social Fields, and Social Boundaries. In *The Archaeology of Social Boundaries*, edited by Miriam T. Stark, pp. 50-77. Smithsonian Institution Press, Washington.

Wertime, Theodore A. 1964. Man's First Encounters with Metallurgy. *Science* 146 (3649): 1257-1267.

Wertime, Theodore A. 1968. A Metallurgical Expedition through the Persian Desert. *Science* l59 (3818): 927-935.

Wertime, Theodore A. 1973a Pyrotechnology: Man's First Industrial Uses of Fire. *American Scientist* 61 (6): 670-682

Wertime, Theodore A. 1973b The Beginnings of Metallurgy. *Science* 182 (4115): 875-887.

Wertime, Theodore A. and James D. Muhly (eds) 1980. *The Coming of Age of Iron*. Yale University Press, New Haven.

Wertime, Theodore A. and Steven F. Wertime (eds) 1982. *Early Pyrotechnology: The Evolution of the First Fire-Using Industries*. Smithsonian Institution Press, Washington.

West, Robert C. 1994. Aboriginal Metallurgy and Metalworking in Spanish America. In *Quest of Mineral Wealth: Aboriginal and Colonial Mining and Metallurgy in Spanish America*, edited by Alan K. Craig and Robert C. West, pp. 5-20. Louisiana State University, Baton Rouge.

White, Leslie A. 1943. Energy and the Evolution of Culture. *American Anthropologist* 45 (3): 335-356.

White, Leslie A. 1949a *The Science of Culture: A Study of Man and Civilization*. Grove Press, New York.

White, Leslie A. 1949b Ethnological Theory. In *Philosophy for the Future: The Quest of Modern Materialism*, edited by Roy W. Sellars, V.J. McGill, and Marvin Farber, Pp. 357-384. Macmillan Co., New York.

White, Leslie A. 1957. Review of Theory of Culture Change: The Methodology of Multilinear Evolution, by Julian H. Steward. *American Anthropologist* 59 (3): 540-542

White, Leslie A. 1959. *The Evolution of Culture: The Development of Civilization to the Fall of Rome*. McGraw-Hill, New York.

Williams, Barbara J. 1972. Tepetate in the Valley of Mexico. *Annals of the Association of American Geographers* 62 (4): 618-626.

Winterhalder, Bruce P., and Eric A. Smith 1992. Evolutionary Ecology and the Social Sciences. In *Evolutionary Ecology and Human Behavior*, edited by

Eric A. Smith, and Bruce P. Winterhalder, pp. 3-23. de Gruyter, New York.

Wittfogel, Karl A. 1955. Developmental aspects of hydraulic societies. In *Irrigation Civilizations: A Comparative Study*, edited by Julian H. Steward, pp. 43-52. Pan American Union Social Science Monographs I, Washington.

Wittfogel, Karl A. 1957. *Oriental Despotism: A Comparative Study of Total Power*. Yale University Press, New Haven.

Yoffee, Norman 1993. Too Many Chiefs? (or, Safe Texts for the '90s). In *Archaeological Theory: Who Sets the Agenda?*, edited by Norman Yoffee and Andrew Sherratt, pp. 60-78. Cambridge University Press, Cambridge.

Zahavi, Amotz and Avishag Zahavi 1997. *The Handicap Principle: A Missing Piece of Darwin's Puzzle*. Oxford University Press, Oxford.

Appendix A

IARP 2003-4: survey data

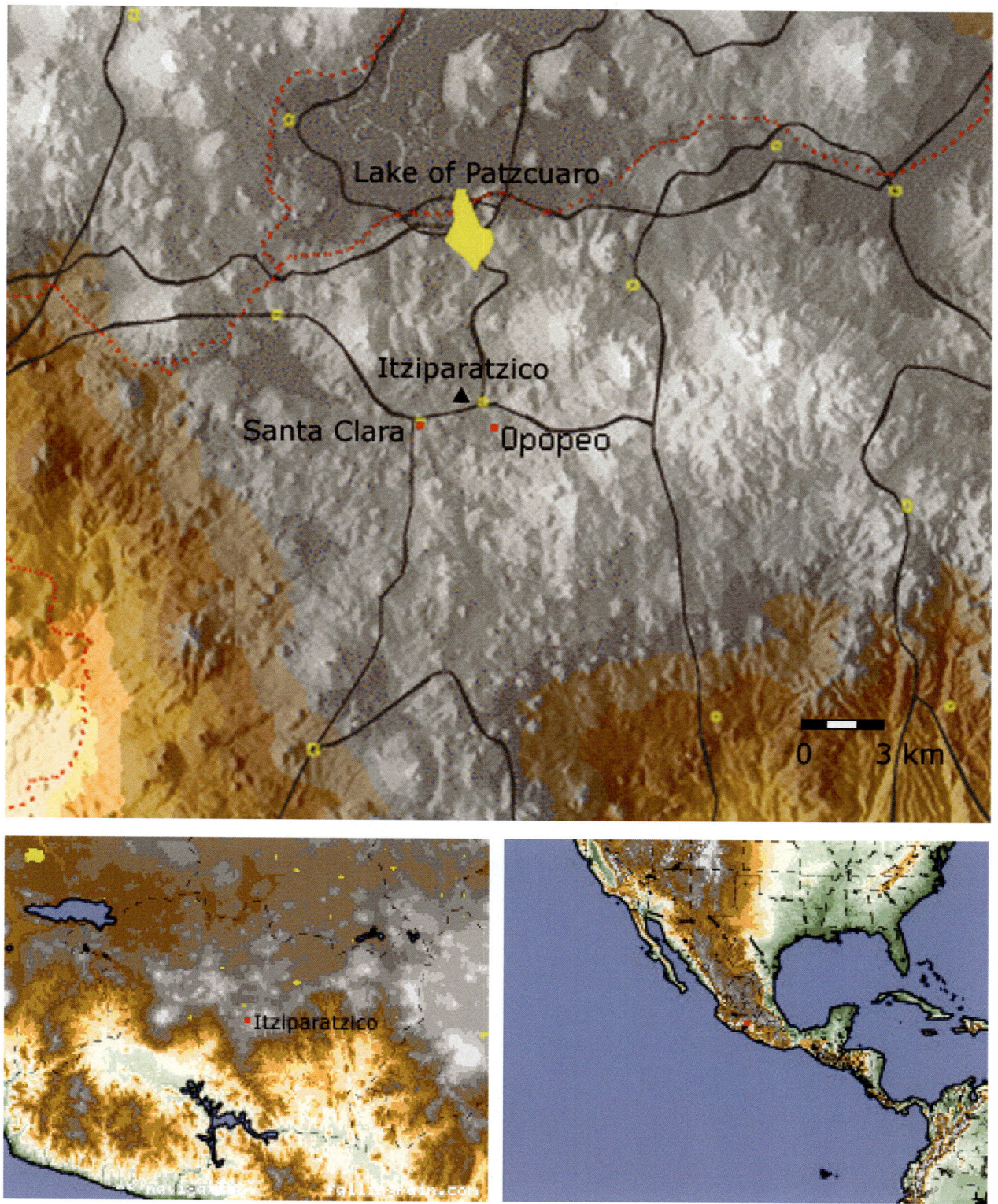

Map A.1 Topographic maps indicating the location of Itziparátzico
(adapted from Opopeo, Mexico Map, Presentation Copyright © Falling Rain Genomics, Inc. 1996-2004).

This Appendix contains information on the features recorded during the IARP 2003 surface survey at Itziparátzico (see Map A.2), as well as an inventory of the archaeological materials collected (Table A.1). Map A.1 shows the survey area in its regional and global context.

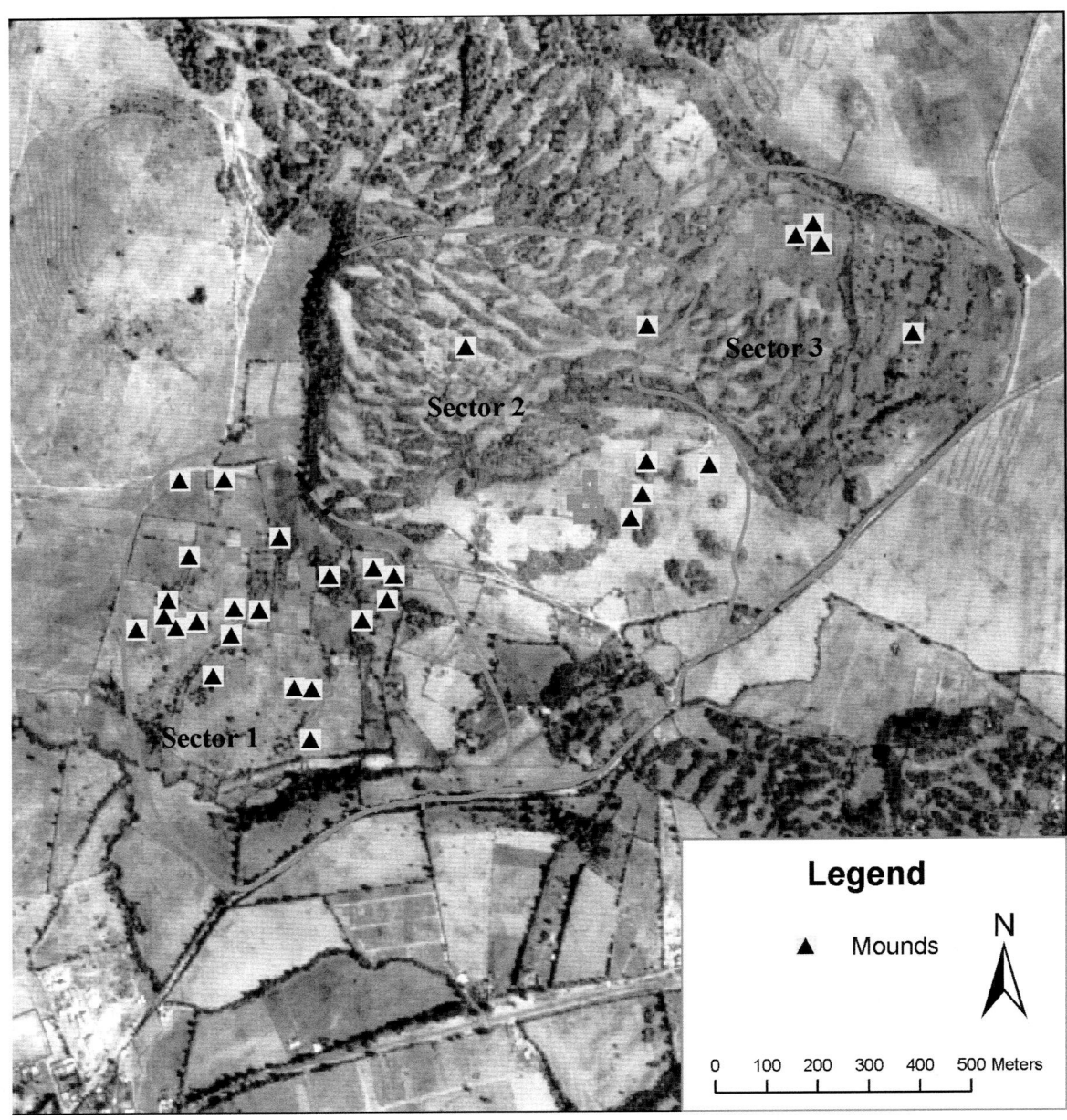

Map A.2 Location of mounds in the surveyed area
(Map created by P. van Rossum, 2006).

Appendix A: IARP 2003-4: Survey Data

Recorded Points and Features by Sector

Sector 1
- Mounds 1-19

Sector 2
- Slope 1
- Mounds 20-24
- Terraces 1-10

Sector 3
- Mounds 25-29
- Terraces 11-27

General Survey
- Fields 1-4
- Mound 31
- Parcels 1-25
- Slag concentrations 1-3

Coordinates of the Surveyed Parcels

Field 1 (x=224171, y=2149423)
Field 2 (x=224319, y=2149655)
Field 3 (x=224120, y=2149271)
Field 4 (x=224330, y=2149525)
Parcel 1 (x=224249, y=2149526)
Parcel 2 (x=224313, y=2149400)
Parcel 3 (x=224239, y=2149261)
Parcel 4 (x=224330, y=2149525)
Parcel 5 (x=224099, y=2149214)
Parcel 6 (x=224093, y=2149157)
Parcel 7 (x=224178, y=2149141)
Parcel 7A (x= 224196, y=2149202)
Parcel 8 (x=224236, y=2149176)
Parcel 9 (x=224231, y=2149099)
Parcel 10 (x=224279, y=2149189)
Parcel 11 (x=224282, y=2149149)
Parcel 12 (x=224343, y=2149060)
Parcel 13 (x=224420, y=2149030)
Parcel 14 (x=224523, y=2149093)
Parcel 15a (x=224505, y=2149111)
Parcel 15b (x=224492, y=2149158)
Parcel 16 (x=224638, y=2149161)
Parcel 17 (x=224629, y=2149080)
Parcel 21 (x=224686, y=2148962)
Parcel 22 (x=224773, y=2149034)
Parcel 23 (x=224036, y=2149084)
Parcel 24 (x=224125, y=2148961)
Parcel 25 (x=224216, y=2148892)

Table A.1 Recovered materials: survey

Bag #	Provenance	Coordinates		Ceramics				Lithics		Slag	Date	Observations
		x	y	Rims	Bodies	Handles	Other	Obsidian	Other			
3	Field 1	224171	2149655	3	39	0	0	0	0	4	13-Jun-03	
6	Field 1			0	2	0	0	0	0	0	13-Jun-03	Majolica (1)
45	Field 1			0	0	0	0	1	0	0	12-Jun-03	Red obsidian frgm
8	Field 2	224319	2149655	0	8	0	0	0	0	0	13-Jun-03	
9	Field 3	224120	2149271	2	34	0	0	0	0	0	13-Jun-03	
1	Field 3			6	13	0	0	27	0	0	13-Jun-03	
5	Parcel 2	224313	2149400	0	0	0	0	1	0	0	13-Jun-03	
7	Parcel 2			1	4	1	1	4	0	0	13-Jun-03	Ring support
2	Parcel 3	224239	2149261	5	40	0	0	0	0	2	13-Jun-03	Polychrome sherds
4	Parcel 3			5	23	0	0	0	0	0	13-Jun-03	Polychrome sherds
48	Parcel 5 (int.)	224099	2149214	5	110	0	0	4	0	7	16-Jun-03	
49	Parcel 5			0	6	0	0	2	0	0	16-Jun-03	
50	Parcel 6 (int.)	224093	2149157	1	53	0	0	2	0	4	16-Jun-03	
51	Parcel 6	224148	2149161	3	0	0	1	9	0	1	16-Jun-03	Ring support
52	Parcel 7 (int.)	224178	2149141	4	54	0	1	7	0	4	16-Jun-03	Ring support
53	Parcel 7a (int.)	224196	2149202	4	86	0	0	4	0	2	16-Jun-03	
54	Parcel 7a			1	5	0	0	0	0	0	16-Jun-03	
55	Parcel 8 (int.)	224236	2149176	3	59	0	0	7	0	30	16-Jun-03	Majolica (1)
56	Parcel 8			0	0	0	0	0	0	0	16-Jun-03	
57	Parcel 10 (int.)	224279	2149189	4	0	0	0	2	0	6	16-Jun-03	
58	Parcel 11 (int.)	224282	2149149	1	8	0	0	0	0	8	16-Jun-03	
59	Parcel 12a (int.)	224343	2149060	1	11	0	0	1	0	223	16-Jun-03	1 out of 3
60	Parcel 12b (int.)	224343	2149060	2	7	1	0	1	0	181	16-Jun-03	2 out of 3
61	Parcel 12c (int.)	224343	2149060	1	9	0	0	2	0	299	16-Jun-03	3 out of 3
77	Parcel 12			1	0	1	0	3	0	0	16-Jun-03	
62	Parcel 13 (int.)	224420	2149030	7	66	0	0	6	0	37	16-Jun-03	
63	Parcel 14 (int.)	224523	2149093	0	22	1	0	3	0	0	17-Jun-03	
64	Parcel 15 (int.)	224505	2149111	0	68	0	0	3	0	28	17-Jun-03	Majolica (4)
79	Parcel 15b (int.)	224492	2149158	0	4	0	0	1	0	1	17-Jun-03	
65	Parcel 23 (int.)	224036	2149084	3	28	0	0	2	0	1	18-Jun-03	
66	Parcel 24 (int.)	224125	2148961	4	14	0	1	0	0	1	18-Jun-03	Conical support
67	Parcel 25 (int.)	224216	2148892	4	32	1	0	6	0	7	18-Jun-03	Majolica (2)
68	Parcel 25			0	8	2	6	31	1	2	18-Jun-03	Tarascan pipes (6)
47	Slag col 3	224295	2149111	0	0	0	0	0	0	64	16-Jun-03	
76	Terrace 27	225347	2149392	0	0	0	0	1	0	0	11-Jun-03	Projectile point
71	Sector 3	225310	2149730	0	0	0	0	1	0	0	11-Jun-03	Projectile point
74	Terrace 26	225232	2149525	0	0	0	0	1	0	0	11-Jun-03	Obsidian scraper
46	Terrace 16	224995	2149565	0	0	0	1	0	0	0	10-Jun-03	Figurine frgmt.
75	Terrace 27			0	0	0	0	1	0	0	11-Jun-03	Red obsidian frgm
69	Out of area			1	3	0	0	3	0	0	9-Jun-03	
70	Out of area			0	0	0	1	5	0	0	9-Jun-03	Tarascan pipe (1)
72	Near M-2			0	4	0	0	0	1	0	9-Jun-03	Basalt scraper
73	Looting M-2			0	1	0	0	0	0	0	9-Jun-03	
	TOTALS			72	821	7	12	141	2	912		

Appendix B

IARP 2003-4: Test pitting data

Seven excavation operations were carried out during the IARP 2003-4 in the three sectors of the Itziparátzico area. Three 2 x 2 m test pits were located in Sector 1, two in Sector 2, and two 2 x 1 m units in Sector 3 (see Map B.1). This appendix describes the excavated units. Also included is an inventory of the archaeological materials recovered from stratigraphic contexts (Table B.1). Although pottery analysis is still pending, an estimated 8, 307 ceramic sherds have been inventoried (Table B.2). All the sherds have been examined by eye and separated into broad categories, i.e. monochromes and polychromes, glazed, and unidentified. Specific ceramic artifacts such as pipes were also quantified. Some of the more frequent wares have been grouped together as 'types'. These include Red-and-White-on-Cream (Red-Wht/Cream), Red with resist decoration (Res/Red) Red-and-Black-on-White (Red-Blk-Wht), and Black on White (Blk/Wht). Examples of these provisional types are provided in Chapter 4 (see Figures 4.7-4.18).

Test Pits Descriptions

Seven units were excavated in arbitrary levels of 20 cm through deposits of silt can clay. There different layers were identified during these operations:

Layer I   A highly organic andosol, locally known as *tupure*. It has a sandy texture and ranges from medium to dark brown.

Layer II   A bright red-colored luvisol locally known as jaboncillo. It has a homogenous, doughy texture and bright red color.

Layer III   *Barro rojo*, a red-colored luvisol with a coarser texture than *jaboncillo*.

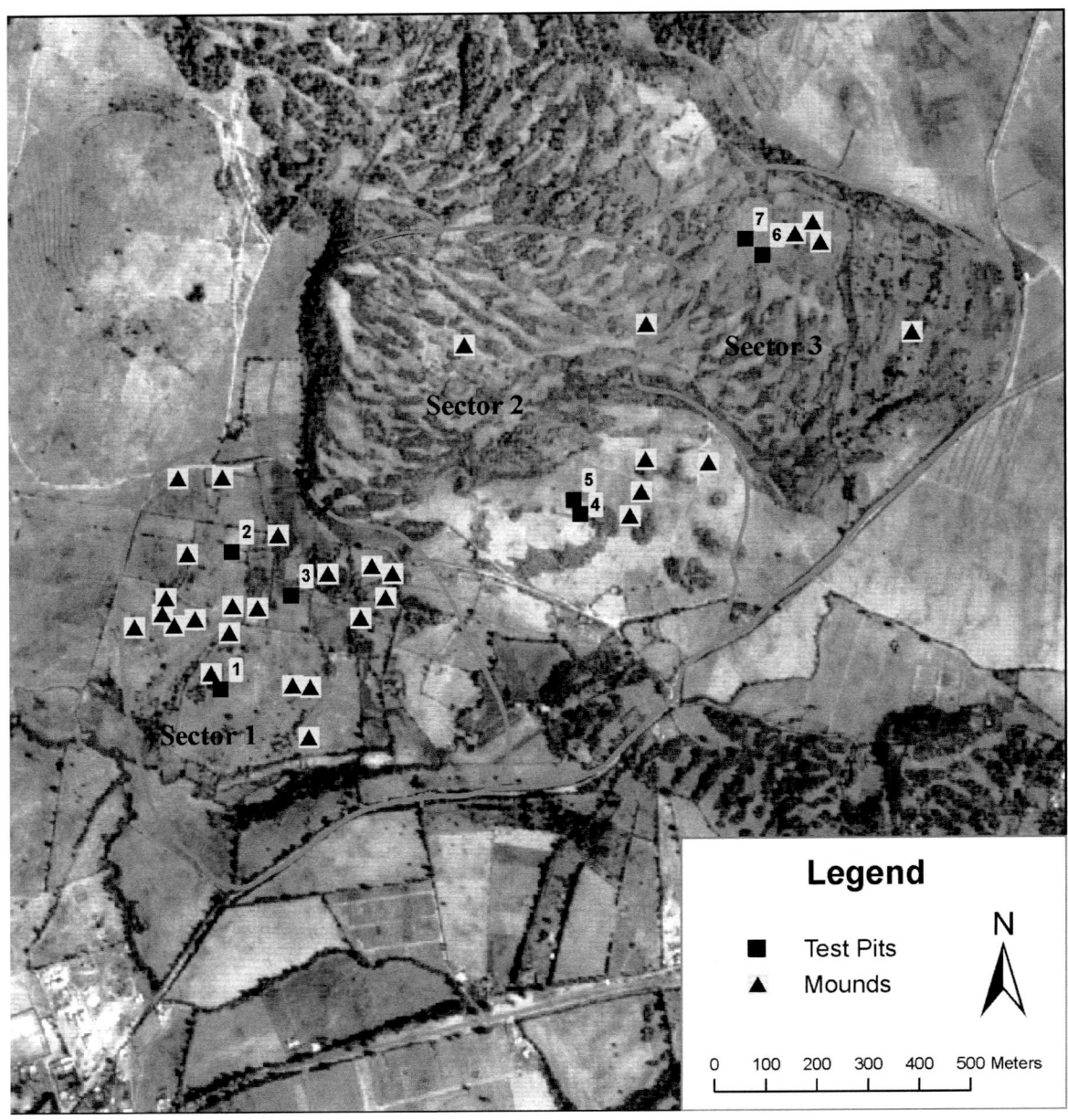

Map B.1 Location of the test pits at Itziparátzico
(Map created by P. van Rossum).

APPENDIX B: IARP 2003-4: TEST PITTING DATA

Unit 1

X= 224206
Y= 2148892 Coordinates on N corner

Unit 1 was a 2x2 m test pit located in Sector 1, in an area with high concentrations of slag. The location was chosen to provide archaeometallurgical samples.

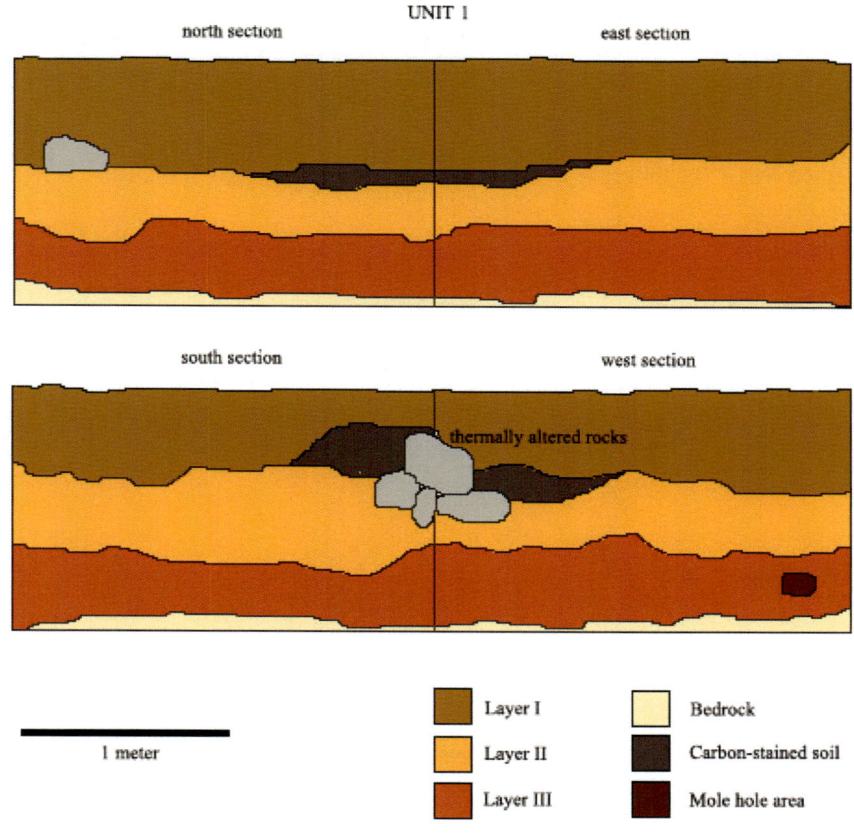

Unit 2

X= 224228
Y= 2149161   SE corner

Unit 2 was a 2x2 m test pit located in Sector 1, in an area with high concentrations of slag. The location was chosen to provide archaeometallurgical samples.

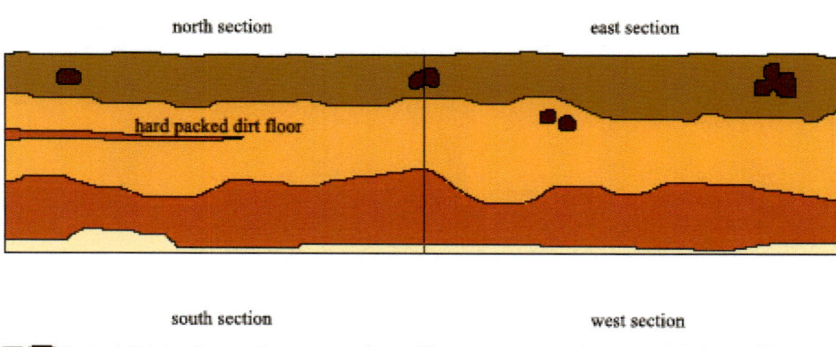

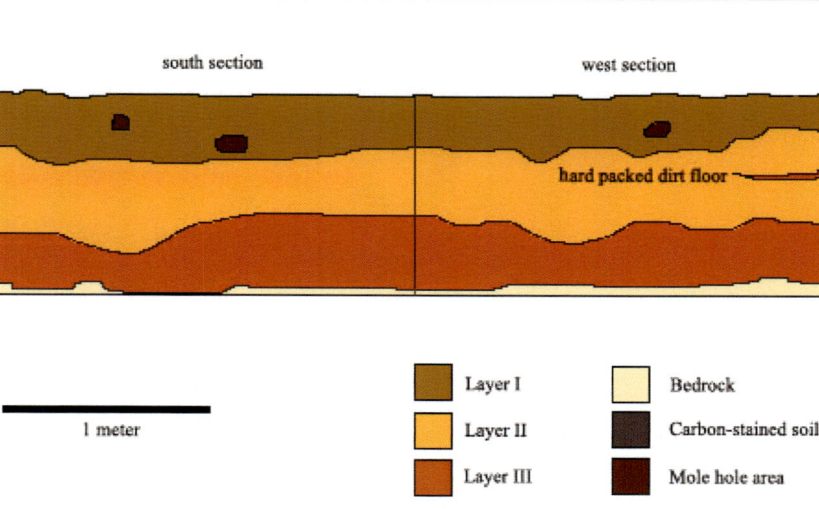

Unit 3

X= 224344
Y= 2149074   N corner

Unit 3 was a 2x2 m test pit located in Sector 1, in an area with high concentrations of slag. The location was chosen to provide archaeometallurgical samples.

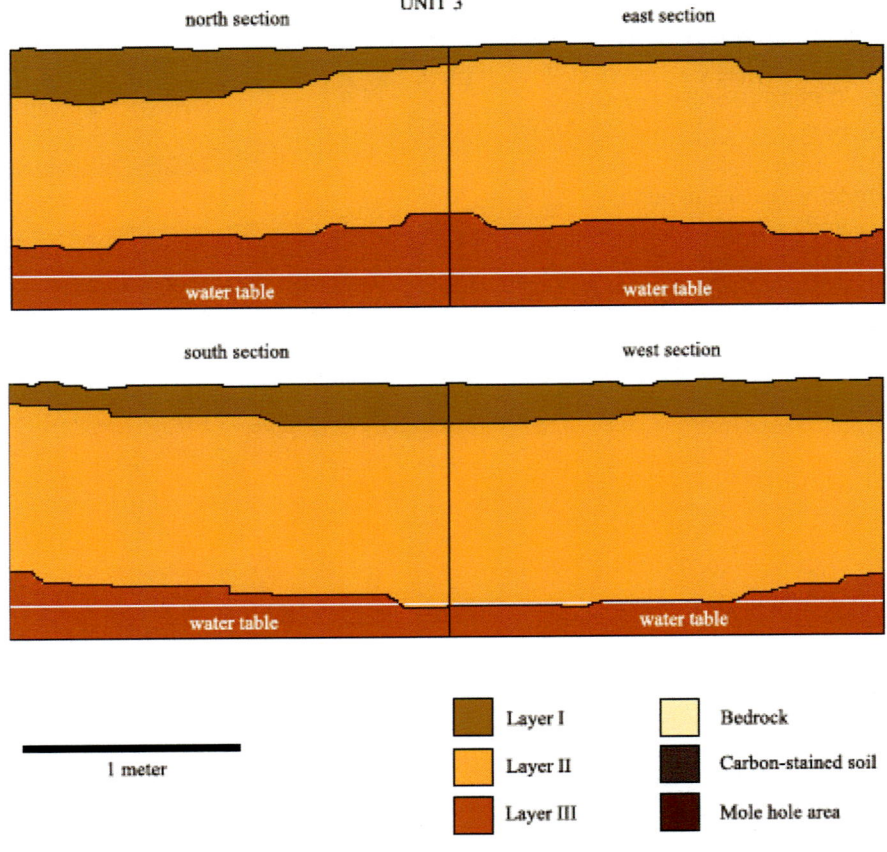

Unit 4

X= 224909
Y= 2149231   SW corner

Unit 4 was a 2x2 m test pit located in Sector 2, on a presumably domestic terrace.

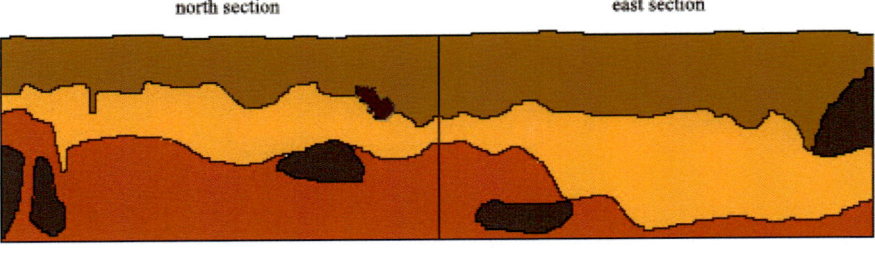

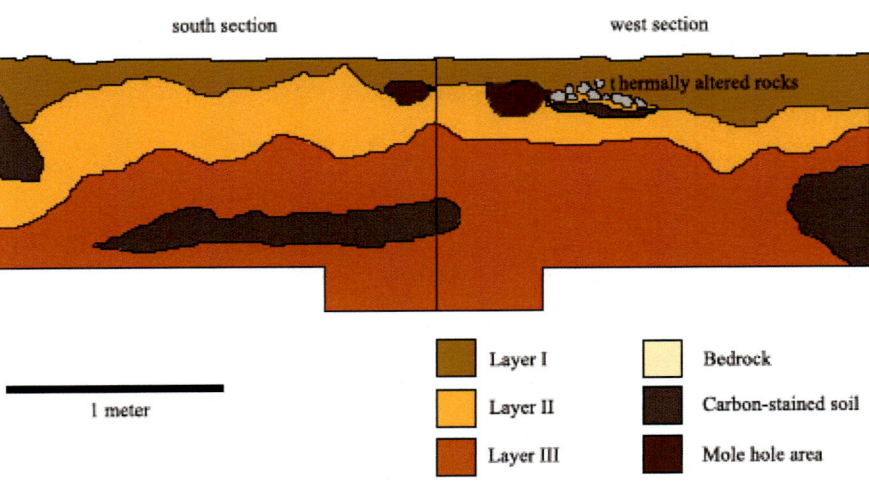

Unit 5

X= 224896
Y= 2149258   SW corner

Unit 5 was a 2x2 m test pit located in Sector 2, on a presumably domestic terrace.

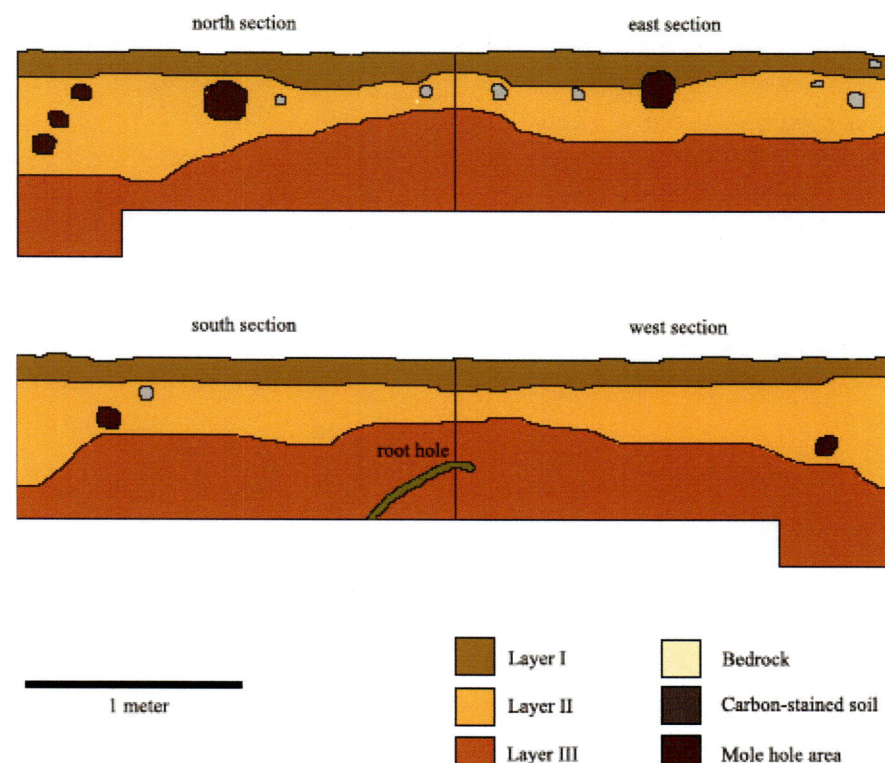

Unit 6

X= 225265
Y= 2149733   SW corner

Unit 6 was a 1x2 m test pit located in Sector 3, in the area of the large mounds. This excavation was complicated by high amounts of large rocks. Considering the proximity of this operation to the structures, it is possible that this area was artificially raised to gain additional building height.

Unit 7

X= 225232
Y= 2149765 SW corner

Unit 7 was a 1x2 m test pit located in Sector 3, in the area of the large mounds. This excavation was complicated by high amounts of large rocks. Considering the proximity of this operation to the structures, it is possible that this area was artificially raised to gain additional building height.

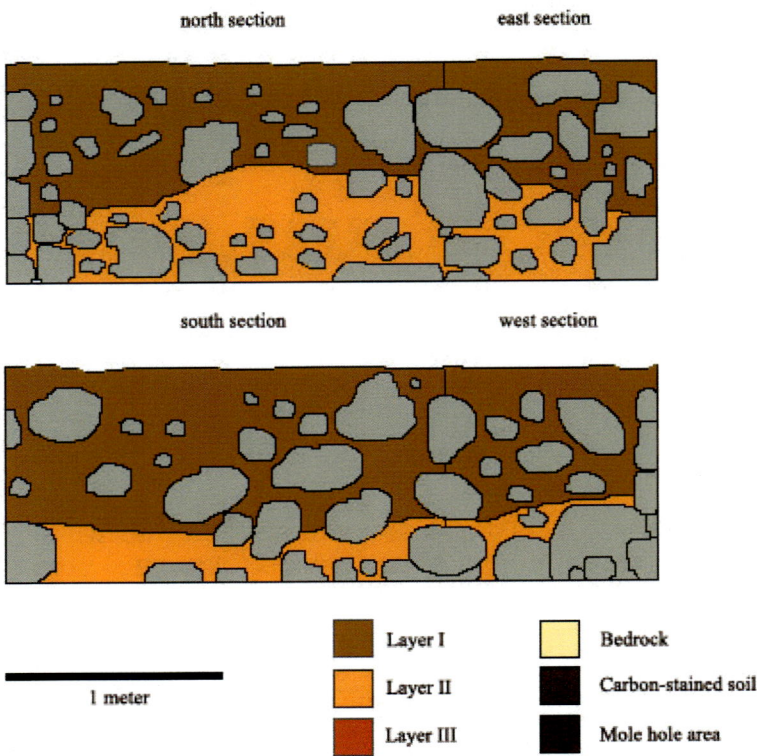

Table B.1 Materials recovered from excavation

Matrix				Ceramics				Lithics			Slag	Other
Sector	Unit	Bag	Level	Monochrome	Polychrome	Unidentified	Glazed	Obsidian	Grinding stone	Other		
1	1	1	1	104	33	252	5	25	0	1	28	0
1	1	2	2	92	25	70	1	8	0	0	32	0
1	1	3	1,2	0	0	0	0	0	0	0	0	1
1	1	4	3	87	20	117	2	11	1	0	19	18
1	1	5	4	110	18	97	2	7	0	2	5	0
1	1	6	5	0	0	0	0	0	0	0	0	1
1	1	7	5	60	13	24	0	3	0	0	0	3
1	1	8	6	23	5	16	0	0	0	0	0	0
1	1	9	7	7	2	10	0	0	0	0	0	0
1	2	1	1	133	67	322	167	14	0	0	24	2
1	2	2	2	0	0	0	0	0	0	0	0	1
1	2	3	2	0	0	0	0	0	0	0	0	1
1	2	4	2	0	0	0	0	0	0	0	0	1
1	2	5	2	31	9	65	2	0	0	0	0	0
1	3	1	1	0	0	13	13	0	0	0	1374	0
1	3	2	1	0	0	19	19	1	0	0	1252	0
1	3	3	1	0	0	6	4	0	0	0	340	0
1	3	4a	1	0	0	0	0	0	0	0	152	0
1	3	4b	1	0	0	32	32	0	0	0	1512	0
1	3	5	1	0	0	23	23	0	0	1	992	0
1	3	6	1	0	0	24	24	1	0	0	1240	0
1	3	7	1	0	0	11	11	2	0	1	949	0
1	3	8	1	0	0	10	10	0	0	0	834	0
1	3	9	1	0	0	28	28	3	0	0	1128	0
1	3	10	1	0	0	15	15	3	0	0	846	0

APPENDIX B: IARP 2003-4: TEST PITTING DATA

Matrix				Ceramics				Lithics			Slag	Other
Sector	Unit	Bag	Level	Monochrome	Polychrome	Unidentified	Glazed	Obsidian	Grinding stone	Other		
1	3	11	1	0	0	48	48	0	0	0	1450	0
1	3	12	1	0	0	30	30	0	0	0	1146	0
1	3	13	2	0	0	0	0	0	0	0	0	1
1	3	14	2	0	0	21	21	0	0	0	514	0
1	3	15	2	0	0	0	0	0	0	0	16	0
1	3	16	3	0	0	0	0	0	0	0	67	0
1	3	17	3	0	0	0	0	0	0	0	0	1
1	3	18	4	0	0	0	0	0	0	0	0	1
1	3	19	4	0	0	9	2	0	0	0	45	0
1	3	20	4	0	0	0	0	0	0	0	0	1
1	3	21	4	0	0	0	0	0	0	0	0	1
1	3	22	4	0	0	0	0	0	0	0	0	1
1	3	23	6	0	0	0	0	0	0	0	0	1
1	3	24	7	0	0	0	0	0	0	0	0	1
1	3	25	8	0	0	0	0	0	0	0	0	0
1	3	26	8	0	0	0	0	0	0	0	0	1
1	3	27	9	0	0	0	0	0	0	0	0	1
1	3	28	9	0	0	0	0	0	0	0	0	1
1	3	29	9	0	0	0	0	0	0	0	0	1
1	3	30	10	0	0	0	0	0	0	0	0	1
2	4	1	1	166	39	849	4	10	0	2	9	1
2	4	2	2	103	17	181	0	2	0	4	0	2
2	4	3	2	0	0	0	0	0	0	0	0	1
2	4	4	3	0	0	0	0	0	0	0	0	1
2	4	5	3	32	10	30	0	3	0	0	0	0
2	4	6	4	0	0	0	0	0	0	0	0	1
2	4	7	4	0	0	0	0	0	0	0	0	1
2	4	8	4	27	9	13	0	1	0	3	0	0
2	4	9	5	4	12	4	0	0	0	0	0	1
2	4	10	5	0	0	0	0	0	0	0	0	1
2	5	1	1	119	76	63	0	9	0	1	1	1
2	5	2	2	123	3	98	0	9	0	1	0	0
2	5	3	2	0	0	0	0	0	0	0	0	1
2	5	4	3	24	0	8	0	2	0	0	0	0
2	5	5	4	0	0	0	0	0	0	0	0	1
2	5	6	4	2	0	4	0	0	0	0	0	0
2	5	7	S	24	2	20	0	2	3	2	0	0
3	6	1	1	48	56	581	0	25	2	43	0	0
3	6	2	2	59	62	545	0	19	0	46	0	0
3	6	3	3	103	47	318	0	5	0	24	0	0
3	6	4	4	0	0	0	0	0	0	0	0	1
3	6	5	4	38	11	87	0	3	0	14	0	1
3	6	6	4	49	6	33	0	1	0	4	0	1
3	7	1	1	87	51	792	1	36	0	59	3	0
3	7	2	2	42	39	171	3	7	0	11	0	1
3	7	3	3	36	20	72	0	1	0	1	0	1
3	7	4	4	19	6	22	0	2	0	6	0	1
3	7	5	5	19	7	25	0	0	0	3	0	0
3	7	6	6	8	1	5	0	1	0	1	0	0
TOTALS				1779	666	5183	467	216	6	230	13978	59

Table B.2 Ceramic inventory

Matrix				Pottery Type										Totals
Sector	Unit	Bag	Level	Red slip	Red-Wht/Cream	Res/Red	Red-Blk-Wht	Blk/Wht	Other Decor	Pipe	Other	Glzd	Un-id	
1	1	1	1	48	5	17	1	1	17	0	33	5	252	379
1	1	2	2	38	7	10	1	0	19	1	44	1	70	191
1	1	3	1,2	0	0	0	0	0	0	0	0	0	0	0
1	1	4	3	41	7	2	1	0	19	0	38	2	117	227
1	1	5	4	50	5	0	0	2	19	0	53	2	97	228
1	1	6	5	0	0	0	0	0	0	0	0	0	0	0
1	1	7	5	16	5	0	0	2	8	0	41	0	24	96
1	1	8	6	7	1	4	0	0	2	0	15	0	16	45
1	1	9	7	6	1	0	0	0	1	0	1	0	11	20
1	2	1	1	53	6	0	0	0	61	2	85	167	322	696
1	2	2	2	0	0	0	0	0	0	0	0	0	0	0
1	2	3	2	0	0	0	0	0	0	0	0	0	0	0
1	2	4	2	0	0	0	0	0	0	0	0	0	0	0
1	2	5	2	12	2	0	0	0	9	0	16	2	65	106
1	3	1	1	0	0	0	0	0	0	0	0	13	32	45
1	3	2	1	0	0	0	0	0	0	0	0	19	45	64
1	3	3	1	0	0	0	0	0	0	0	0	4	6	10
1	3	4a	1	0	0	0	0	0	0	0	0	0	1	1
1	3	4b	1	0	0	0	0	0	0	0	0	32	66	98
1	3	5	1	0	0	0	0	0	0	0	0	23	36	59
1	3	6	1	0	0	0	0	0	0	0	0	24	36	60
1	3	7	1	0	0	0	0	0	0	0	0	11	45	56
1	3	8	1	0	0	0	0	0	0	0	0	10	26	36
1	3	9	1	0	0	0	0	0	0	0	0	28	36	64
1	3	10	1	0	0	0	0	0	0	0	0	15	31	46
1	3	11	1	0	0	0	0	0	0	0	0	48	79	127
1	3	12	1	0	0	0	0	0	0	0	0	30	64	94
1	3	13	2	0	0	0	0	0	0	0	0	0	0	0
1	3	14	2	0	0	0	0	0	0	0	0	21	30	51
1	3	15	2	0	0	0	0	0	0	0	0	0	1	1
1	3	16	3	0	0	0	0	0	0	0	0	0	3	3
1	3	17	3	0	0	0	0	0	0	0	0	0	0	0
1	3	18	4	0	0	0	0	0	0	0	0	0	0	0
1	3	19	4	0	0	0	0	0	0	0	0	2	9	11
1	3	20	4	0	0	0	0	0	0	0	0	0	0	0
1	3	21	4	0	0	0	0	0	0	0	0	0	0	0
1	3	22	4	0	0	0	0	0	0	0	0	0	0	0
1	3	23	6	0	0	0	0	0	0	0	0	0	0	0
1	3	24	7	0	0	0	0	0	0	0	0	0	0	0
1	3	25	8	0	0	0	0	0	0	0	0	0	0	0
1	3	26	8	0	0	0	0	0	0	0	0	0	0	0
1	3	27	9	0	0	0	0	0	0	0	0	0	0	0
1	3	28	9	0	0	0	0	0	0	0	0	0	0	0
1	3	29	9	0	0	0	0	0	0	0	0	0	0	0
1	3	30	10	0	0	0	0	0	0	0	0	0	0	0
2	4	1	1	92	7	4	4	0	24	0	67	4	849	1051
2	4	2	2	35	3	1	2	0	21	0	58	0	181	301
2	4	3	2	0	0	0	0	0	0	0	0	0	0	0

Appendix B: IARP 2003-4: Test pitting data

Matrix				Pottery Type										
Sector	Unit	Bag	Level	Red slip	Red-Wht/Cream	Res/Red	Red-Blk-Wht	Blk/Wht	Other Decor	Pipe	Other	Glzd	Un-id	Totals
2	4	4	3	0	0	0	0	0	0	0	0	0	0	0
2	4	5	3	5	1	2	4	0	19	0	11	0	30	72
2	4	6	4	0	0	0	0	0	0	0	0	0	0	0
2	4	7	4	0	0	0	0	0	0	0	0	0	0	0
2	4	8	4	3	2	1	0	1	7	0	26	0	13	53
2	4	9	5	0	2	0	0	0	11	0	3	0	4	20
2	4	10	5	0	0	0	0	0	0	0	0	0	0	0
2	2	11	S	9	0	0	2	1	2	1	5	0	8	28
2	5	1	1	30	0	0	0	0	8	2	86	0	63	189
2	5	2	2	42	13	1	0	0	4	1	65	0	98	224
2	5	3	2	0	0	0	0	0	0	0	0	0	0	0
2	5	4	3	3	0	0	0	0	0	0	21	0	8	32
2	5	5	4	0	0	0	0	0	0	0	0	0	0	0
2	5	6	4	1	0	0	0	0	0	0	1	0	4	6
2	5	7	S	16	0	0	0	0	2	2	4	0	20	44
3	6	1	1	22	1	0	1	1	54	0	24	0	581	684
3	6	2	2	19	2	0	1	2	64	2	30	0	545	665
3	6	3	3	15	1	0	2	1	67	0	64	0	318	468
3	6	4	4	0	0	0	0	0	0	0	0	0	0	0
3	6	5	4	5	0	0	0	0	16	0	28	0	87	136
3	6	6	4	0	18	5	0	1	3	0	28	0	33	88
3	7	1	1	35	10	1	0	0	48	1	41	3	792	931
3	7	2	2	7	3	4	1	1	34	0	32	1	171	254
3	7	3	3	6	6	1	1	0	27	0	15	0	72	128
3	7	4	4	0	0	0	2	0	11	0	19	0	22	54
3	7	5	5	2	2	0	0	1	6	0	15	0	25	51
3	7	6	6	0	0	0	0	0	0	0	9	0	5	14
	Totals			618	110	53	23	14	583	12	978	467	5449	8307

Appendix C

Slag analyses

Copper smelting slag recovered from the excavations at Itziparátzico was analyzed for microstructure and compositional properties using light microscopy, scanning electron microscopy with energy-dispersive x-ray spectrometry (SEM/EDS), and x-ray fluorescence spectrometry (XRF). This appendix presents the results of these analyses, as well as a brief explanation of phase diagrams and their use in archaeometallurgical studies.

The samples selected for analysis were recovered from units 1, 2, 3, 4, and 7, and were associated with Late Postclassic artifacts. Three samples from Jicalán El Viejo were also included. Polished and thin sections were prepared for the following specimens:

Abbreviation	Provenance		Classification	SEM Data Yes/No
	Exc. Unit	Strat. Level	Platy/Lumpy	
1-1a	1	1	Platy (p)	Y
1-1b	1	1	Platy	Y
1-2b	1	2	Platy	Y
1-4a	1	4	Platy	Y
2-1b	2	1	Platy	Y
4-1a	4	1	Platy	N
5-1a	5	1	Platy	N
7-1a	7	1	Platy	N
7-1b	7	1	Lumpy/Platy (p/l)	N
1-2a	1	2	Lumpy (l)	Y
1-3a	1	3	Lumpy	Y
1-3b	1	3	Lumpy	Y
2-1a	2	1	Lumpy	Y
3-1a	3	1	Lumpy	N
3-1b	3	1	Lumpy	N
3-1c	3	1	Lumpy	N
J-sa	N/A -Jicalán	Surface	Lumpy	N
J-sb	N/A -Jicalán	Surface	Lumpy	N
J-sc	N/A -Jicalán	Surface	Platy	N

Sample 1-1a: Light Microscopy and SEM/EDS Analyses

Polished thin section

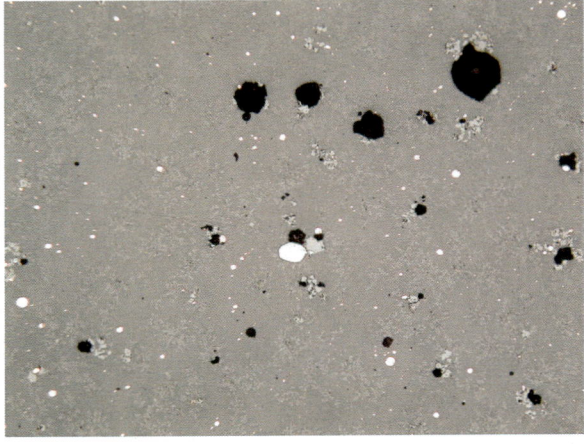

Micrograph showing magnetite crystals surrounded by a glassy matrix rich in iron oxide, sulfide inclusions, and a copper metal prill. x 62.5.

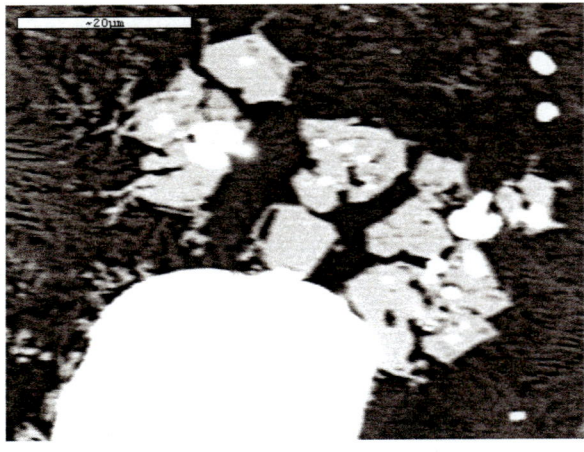

SEM micrograph of copper metal prill and associated magnetite crystals.

Appendix C: Slag Analyses

Sample 1-1a: SEM analysis of copper prill.

SEMQuant results. Listed at 2:31:18 PM on 8/10/04
Operator: blanca maldonado
Client: none
Job: blanca maldonado
Spectrum label: PSC-04-1/I-1A-clstr3/prll-1

System resolution = 83 eV

Quantitative method: ZAF (2 iterations).
Analysed all elements and normalised results.

5 peaks possibly omitted: 0.00, 0.26, 0.50, 0.92, 1.82 keV

Standards :
  S   K   sk 12/13/01
  Fe   K   fek 5/19/04
  Ni   K   nil 5/19/04
  Cu   K   cul 5/19/04
  Zn   K   znk 5/18/04
  As   K   ask 12/13/01
  Sn   L   snl 5/19/04
  Pb   L   Pbl 11/11/98

Elmt	Spect. Type	Element %	Atomic %
S K	ED	0.06	0.12
Fe K	ED	2.37	2.69
Ni K	ED	0.02*	0.02*
Cu K	ED	97.29	97.20
Zn K	ED	-0.08*	-0.08*
As K	ED	-0.13*	-0.11*
Sn L	ED	0.01*	0.01*
Pb L	ED	0.45*	0.14*
Total		100.00	100.00

* = <2 Sigma

Sample 1-1a: SEM analysis of magnetite crystals.

SEMQuant results. Listed at 2:12:24 PM on 8/10/04
Operator: blanca maldonado
Client: none
Job: blanca maldonado
Spectrum label: PSC-04-1/I-1A-clstr2/Fe3O4-2

System resolution = 83 eV

Quantitative method: ZAF (2 iterations).
Analysed elements combined with: O (Valency: -2)
Method : Stoichiometry Normalised results.
Nos. of ions calculation based on 32 anions per formula.

1 peak possibly omitted: 0.00 keV

Standards :
  Na   K   NAK 6/11/04
  Mg   K   MgO 01/12/93
  Al   K   alk 5/26/04
  Si   K   siwol 5/25/04
  P   K   GaP 29/11/93
  S   K   sk 12/13/01
  K   K   KK 6/11/04
  Ca   K   cak 5/25/04
  Ti   K   Ti 01/12/93
  Mn   K   mnk 12/14/01
  Fe   K   fek 5/19/04
  Ni   K   nil 5/19/04
  Cu   K   cul 5/19/04
  Zn   K   znk 5/18/04
  Pb   L   Pbl 11/11/98

Elmt	Spect. Type	Element %	Atomic %	Compound	Nos. of ions
Na K	ED	0.79	1.07	Na2O	1.06	0.65
Mg K	ED	0.49	0.63	MgO	0.81	0.38
Al K	ED	8.13	9.46	Al2O3	15.36	5.77
Si K	ED	0.26	0.30	SiO2	0.56	0.18
P K	ED	0.01*	0.01*	P2O5	0.02*	0.01*
S K	ED	0.00*	0.00*	SO3	0.00*	0.00*
K K	ED	0.00*	0.00*	K2O	0.00*	0.00*
Ca K	ED	0.04*	0.03*	CaO	0.05*	0.02*
Ti K	ED	0.73	0.48	TiO2	1.23	0.29
Mn K	ED	0.03*	0.02*	MnO	0.04*	0.01*
Fe K	ED	62.98	35.42	FeO	81.02	21.59
Ni K	ED	0.01*	0.01*	NiO	0.02*	0.00*

```
Cu K      ED       0.28      0.14    CuO    0.35    0.08
Zn K      ED       0.05*     0.02*   ZnO    0.06*   0.01*
Pb L      ED      -0.54*    -0.08*   PbO   -0.58*  -0.05*
O                 26.74     52.49          32.00
Total            100.00    100.00         100.00

Cation sum 28.96
* = <2 Sigma
```

Sample 1-1a: SEM/EDS analysis

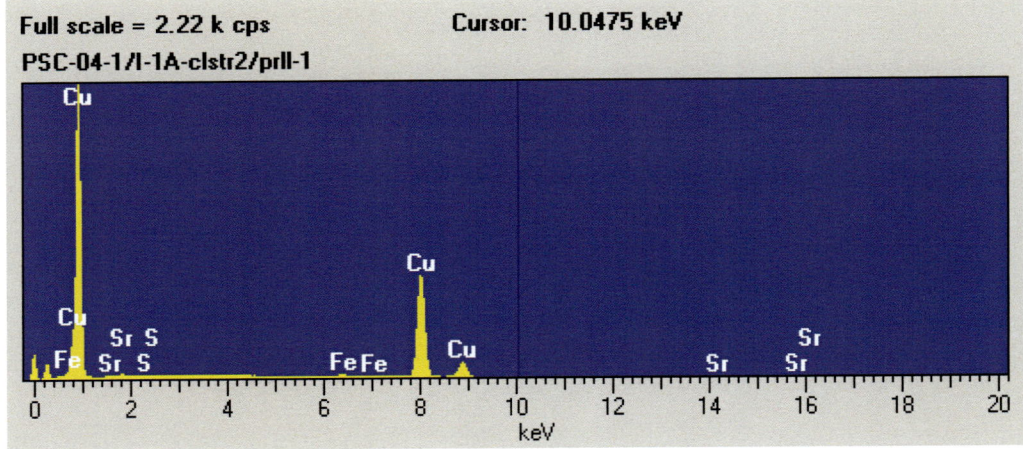

SEM analysis of copper prill.

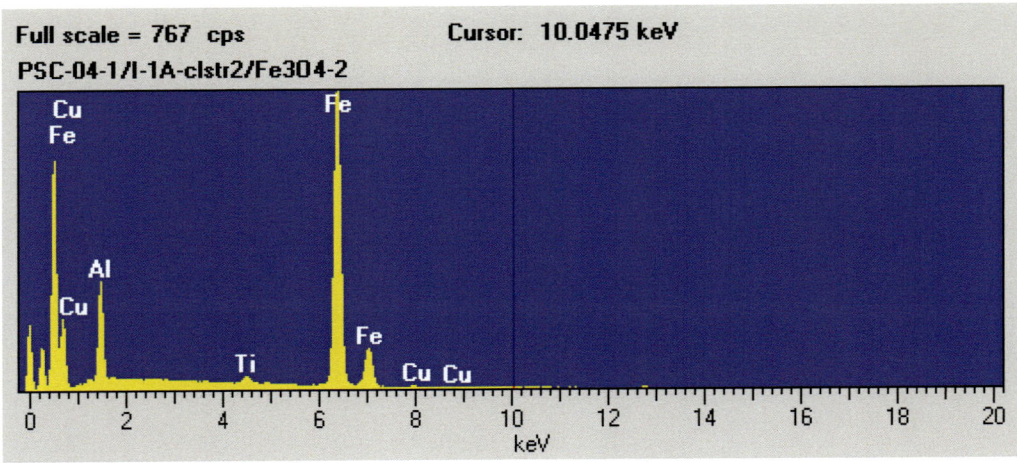

SEM analysis of magnetite crystals.

Sample 1-1b: Light Microscopy and SEM/EDS Analyses

Polished thin section

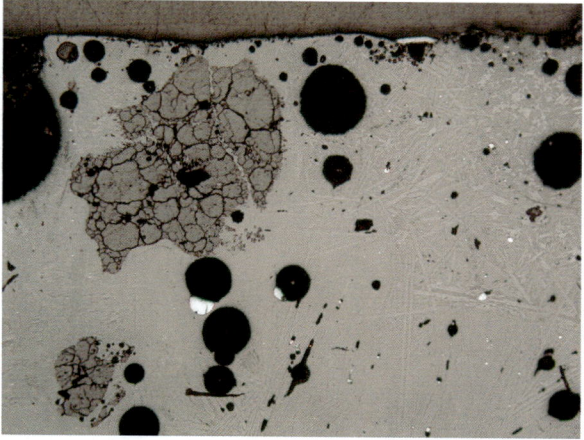

Micrograph showing fayalite crystals surrounded by a glassy matrix rich in iron oxide, sulfide inclusions, together with a polycrystalline aggregate of un-reacted quartz. x 62.5.

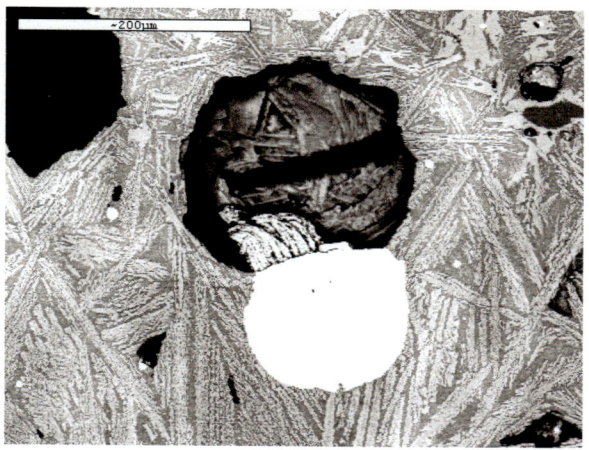

SEM micrograph of copper sulfide. The network of fayalitic plates (light gray) is typical of eutectic solidification.

Sample 1-1b: SEM Analysis of Copper Sulfide

SEMQuant results. Listed at 2:56:13 PM on 8/10/04
Operator: blanca maldonado
Client: none
Job: blanca maldonado
Spectrum label: PSC-04-1/I-1B-clstr2/CuS-1

System resolution = 82 eV

Quantitative method: ZAF (2 iterations).
Analysed all elements and normalised results.

5 peaks possibly omitted: 0.00, 0.26, 0.52,
0.92, 1.80 keV

Standards :
S K	sk	12/13/01
Fe K	fek	5/19/04
Ni K	nil	5/19/04
Cu K	cul	5/19/04
Zn K	znk	5/18/04
As K	ask	12/13/01
Sn L	snl	5/19/04
Pb L	Pbl	11/11/98

Elmt	Spect. Type	Element %	Atomic %
S K	ED	19.69	32.66
Fe K	ED	1.11	1.06
Ni K	ED	0.06*	0.05*
Cu K	ED	79.50	66.56
Zn K	ED	-0.04*	-0.03*
As K	ED	-0.42*	-0.30*
Sn L	ED	-0.16*	-0.07*
Pb L	ED	0.27*	0.07*
Total		100.00	100.00

* = <2 Sigma

Tarascan Copper Metallurgy

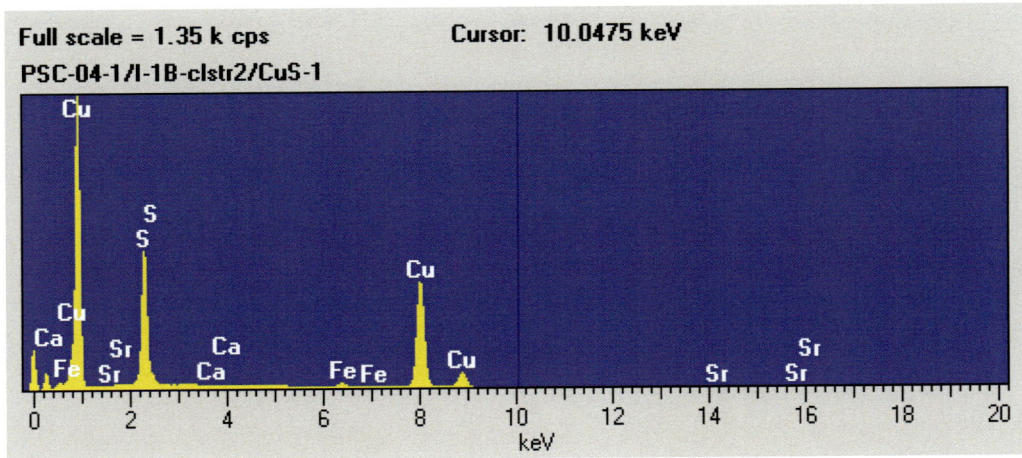

Sample 1-2b: Light Microscopy and SEM/EDS Analyses

Polished thin section

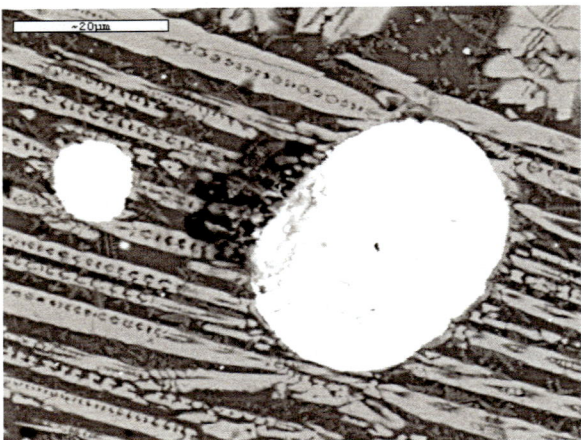

SEM micrograph of copper sulfides and associated fayalite crystals.

Micrograph showing sulfide inclusions in a glassy matrix. x 62.5.

Sample 1-2b: SEM Analysis of Copper Sulfide

SEMQuant results. Listed at 3:18:44 PM on 8/10/04
Operator: blanca maldonado
Client: none
Job: blanca maldonado
Spectrum label: PSC-04-1/II-2B-clstr1/CuS-1

System resolution = 81 eV

Quantitative method: ZAF (2 iterations).
Analysed all elements and normalised results.

7 peaks possibly omitted: 0.00, 0.26, 0.52, 0.92, 1.48, 1.82, 16.94 keV

APPENDIX C: SLAG ANALYSES

```
Standards :
S   K      sk 12/13/01
Fe  K      fek 5/19/04
Ni  K      nil 5/19/04
Cu  K      cul 5/19/04
Zn  K      znk 5/18/04
As  K      ask 12/13/01
Sn  L      snl 5/19/04
Pb  L      Pbl 11/11/98
```

Elmt	Spect. Type	Element %	Atomic %
S K	ED	19.63	32.70
Fe K	ED	1.34	1.28
Ni K	ED	0.00*	0.00*
Cu K	ED	78.87	66.27
Zn K	ED	0.10*	0.08*
As K	ED	-0.67*	-0.48*
Sn L	ED	-0.20*	-0.09*
Pb L	ED	0.93*	0.24*
Total		100.00	100.00

* = <2 Sigma

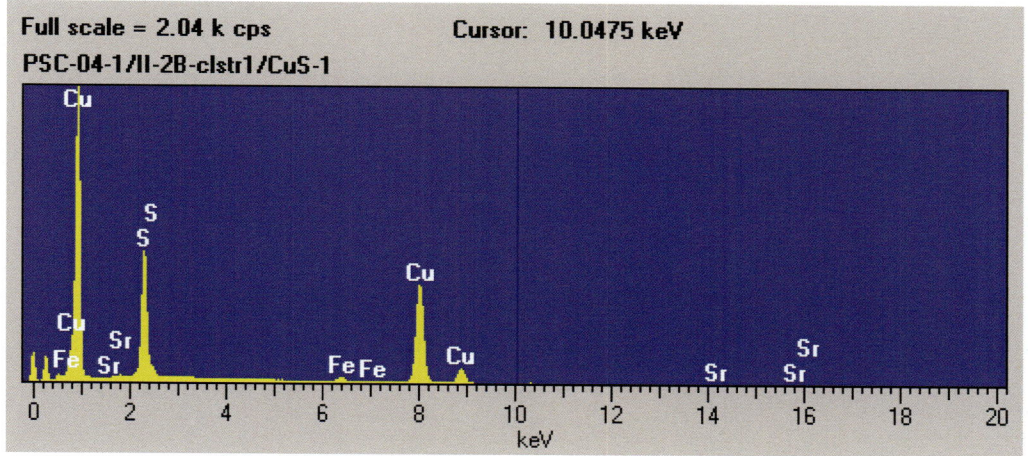

Sample 1-4a: Light Microscopy and SEM/EDS Analyses

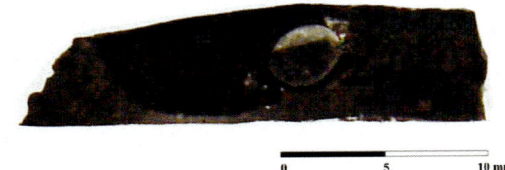

Polished thin section.

Micrograph showing duplex copper/copper sulfide prills surrounded by a glassy matrix. x 62.5.

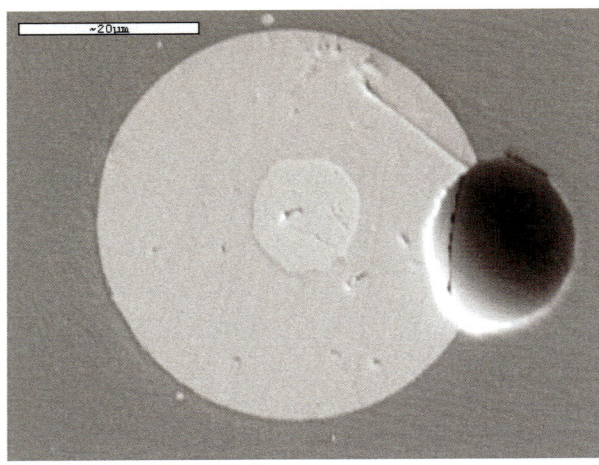

SEM micrograph showing duplex copper/copper sulfide prills surrounded by a glassy matrix.

Sample 1-4a: SEM analysis of unresolved eutectic mixture.

SEMQuant results. Listed at 2:00:18 PM on 8/16/04
Operator: blanca maldonado
Client: none
Job: blanca maldonado
Spectrum label: PSC-04-1/I-4A-mtrx-1

System resolution = 82 eV

Quantitative method: ZAF (3 iterations).
 Analysed elements combined with: O (Valency: -2)
 Method : Stoichiometry Normalised results.
 Nos. of ions calculation based on 32 anions per formula.

1 peak possibly omitted: 0.00 keV

Standards :
Element	Standard
Na K	NAK 6/11/04
Mg K	MgO 01/12/93
Al K	alk 5/26/04
Si K	siwol 5/25/04
P K	GaP 29/11/93
S K	sk 12/13/01
K K	KK 6/11/04
Ca K	cak 5/25/04
Ti K	Ti 01/12/93
Mn K	mnk 12/14/01
Fe K	fek 5/19/04
Ni K	nil 5/19/04
Cu K	cul 5/19/04
Zn K	znk 5/18/04
Pb L	Pbl 11/11/98

Elmt	Spect. Type	Element %	Atomic %	Compound	Compound %	Nos. of ions
Na K	ED	0.64	0.70	Na2O	0.86	0.37
Mg K	ED	1.19	1.23	MgO	1.98	0.66
Al K	ED	6.72	6.25	Al2O3	12.70	3.37
Si K	ED	17.83	15.91	SiO2	38.14	8.58
P K	ED	0.01*	0.01*	P2O5	0.03*	0.01*
S K	ED	0.19	0.15	SO3	0.47	0.08
K K	ED	1.18	0.75	K2O		

APPENDIX C: SLAG ANALYSES

```
1.42          0.41                           System resolution =  82 eV
Ca K     ED      1.41      0.88    CaO
1.97          0.47                           Quantitative method: ZAF (3
Ti K     ED      0.24      0.13    TiO2     iterations).
0.40          0.07                            Analysed all elements and normalised
Mn K     ED      0.05*     0.02*   MnO      results.
0.07*         0.01*
Fe K     ED     32.24     14.47    FeO      5 peaks possibly omitted: 0.00,
41.48         7.80                           0.26, 0.54,
Ni K     ED      0.00*     0.00*   NiO      0.92, 1.80 keV
0.00*         0.00*
Cu K     ED      0.35      0.14    CuO     Standards :
0.44          0.08                            S   K       sk 12/13/01
Zn K     ED     -0.06*    -0.02*   ZnO       Fe  K       fek 5/19/04
-0.07*        -0.01*                          Ni  K       nil 5/19/04
Pb L     ED      0.11*     0.01*   PbO       Cu  K       cul 5/19/04
0.12*         0.01*                           Zn  K       znk 5/18/04
O               37.89     59.37               As  K       ask 12/13/01
32.00                                         Sn  L       snl 5/19/04
Total          100.00    100.00               Pb  L       Pbl 11/11/98
100.00
```

Cation sum 21.90
* = <2 Sigma

Sample 1-4a: SEM analysis of duplex copper/copper sulfide prills.

SEMQuant results. Listed at 3:58:17 PM on 8/16/04
Operator: blanca maldonado
Client: none
Job: blanca maldonado
Spectrum label: PSC-04-1/II-4A-prll3/xt

```
Elmt     Spect.   Element   Atomic
         Type        %         %
S   K    ED       20.80     34.08
Fe  K    ED        1.32      1.24
Ni  K    ED        0.04*     0.04*
Cu  K    ED       78.54     64.95
Zn  K    ED        0.00*     0.00*
As  K    ED       -0.26*    -0.19*
Sn  L    ED       -0.04*    -0.02*
Pb  L    ED       -0.39*    -0.10*
Total            100.00    100.00
```

* = <2 Sigma

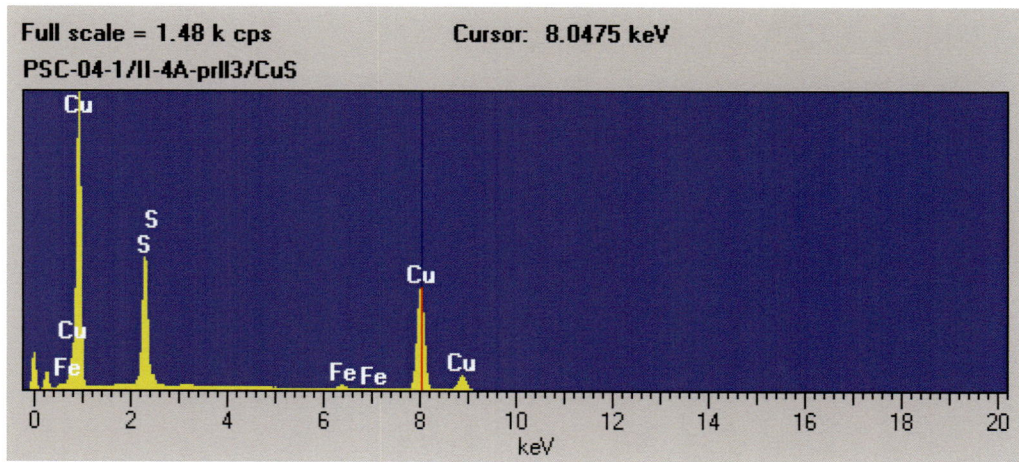

Sample 2-1b: Light Microscopy and SEM/EDS Analyses

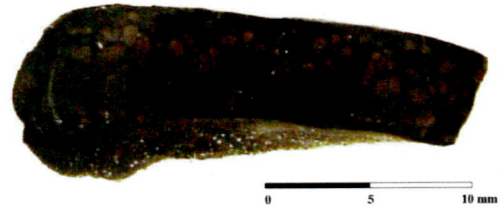

Polished thin section.

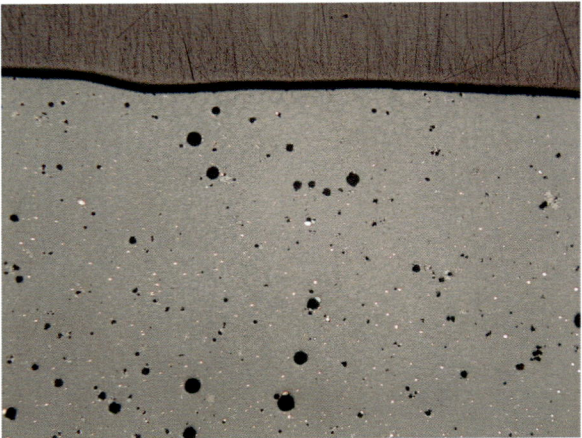

Micrograph of a typical region of a platy slag. x 62.5.

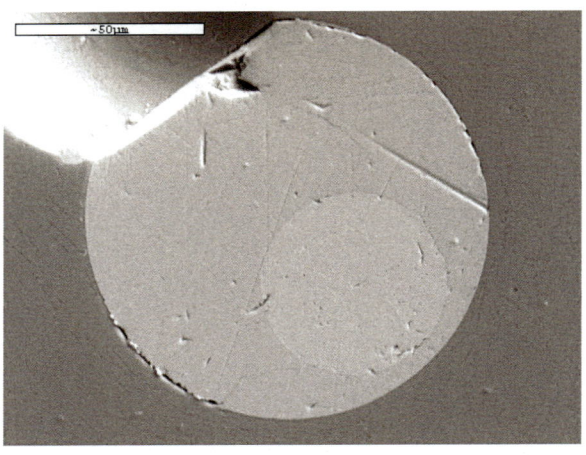

SEM micrograph showing duplex copper/copper sulfide prills surrounded by a glassy matrix.

Sample 2-1b: SEM analysis of duplex copper/copper sulfide prills.

SEMQuant results. Listed at 3:18:44 PM on 8/10/04
Operator: blanca maldonado
Client: none
Job: blanca maldonado
Spectrum label: PSC-04-1/II-2B-clstr1/CuS-1

System resolution = 81 eV

Quantitative method: ZAF (2 iterations).
 Analysed all elements and normalised results.

7 peaks possibly omitted: 0.00, 0.26, 0.52,
 0.92, 1.48, 1.82, 16.94 keV

Standards :
S   K   sk 12/13/01
Fe K   fek 5/19/04
Ni K   nil 5/19/04
Cu K   cul 5/19/04
Zn K   znk 5/18/04
As K   ask 12/13/01
Sn L   snl 5/19/04
Pb L   Pbl 11/11/98

Elmt	Spect. Type	Element %	Atomic %
S K	ED	19.63	32.70
Fe K	ED	1.34	1.28
Ni K	ED	0.00*	0.00*
Cu K	ED	78.87	66.27
Zn K	ED	0.10*	0.08*
As K	ED	-0.67*	-0.48*
Sn L	ED	-0.20*	-0.09*
Pb L	ED	0.93*	0.24*
Total		100.00	100.00

* = <2 Sigma

APPENDIX C: SLAG ANALYSES

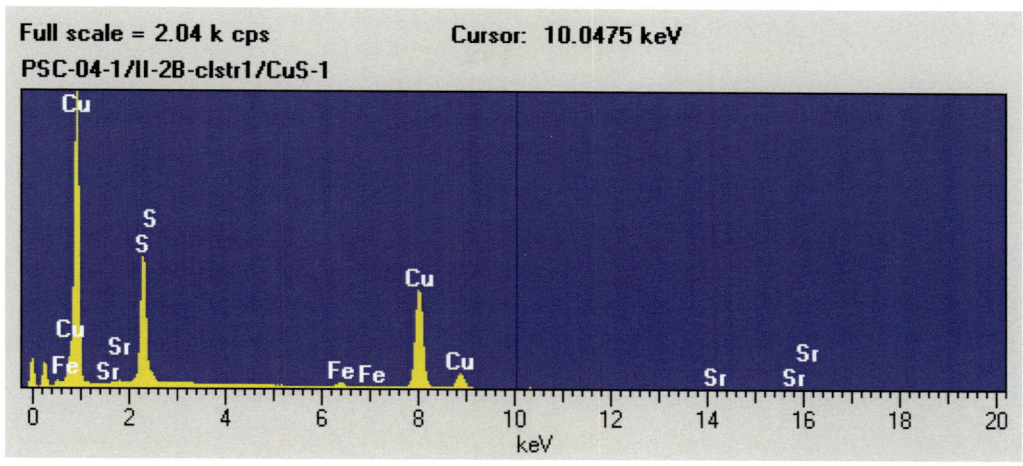

Sample 2-1b: SEM analysis of duplex copper/copper sulfide prills.

SEMQuant results. Listed at 3:20:04 PM on 8/10/04
Operator: blanca maldonado
Client: none
Job: blanca maldonado
Spectrum label: PSC-04-1/II-2B-clstr1/CuS-2

System resolution = 81 eV

Quantitative method: ZAF (2 iterations).
 Analysed all elements and normalised results.

 6 peaks possibly omitted: 0.00, 0.26, 0.52,
 0.92, 1.82, 16.80 keV

Standards :
S K	sk	12/13/01
Fe K	fek	5/19/04
Ni K	nil	5/19/04
Cu K	cul	5/19/04
Zn K	znk	5/18/04
As K	ask	12/13/01
Sn L	snl	5/19/04
Pb L	Pbl	11/11/98

Elmt	Spect. Type	Element %	Atomic %
S K	ED	0.21	0.42
Fe K	ED	3.45	3.90
Ni K	ED	0.10*	0.10*
Cu K	ED	95.02	94.59
Zn K	ED	0.16*	0.15*
As K	ED	1.00	0.85
Sn L	ED	-0.16*	-0.08*
Pb L	ED	0.22*	0.07*
Total		100.00	100.00

* = <2 Sigma

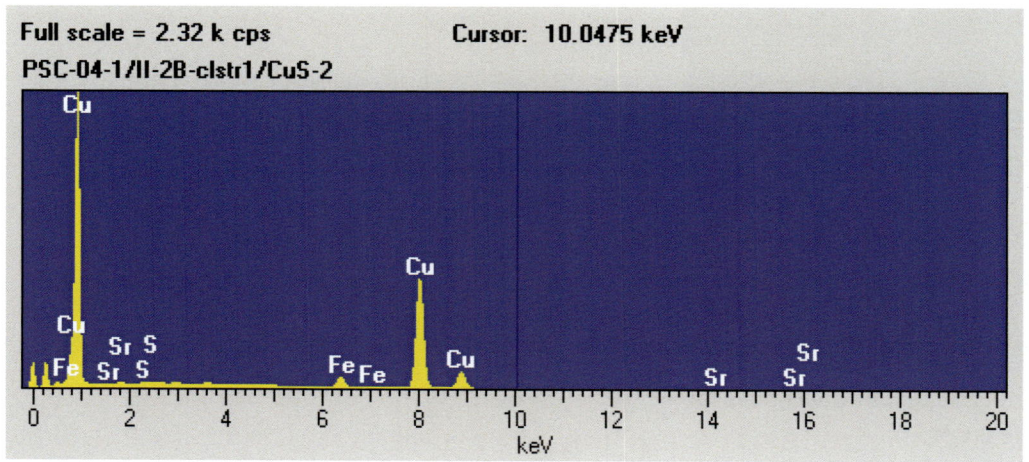

Sample 1-2a: Light Microscopy and SEM/EDS Analyses

Polished thin section.

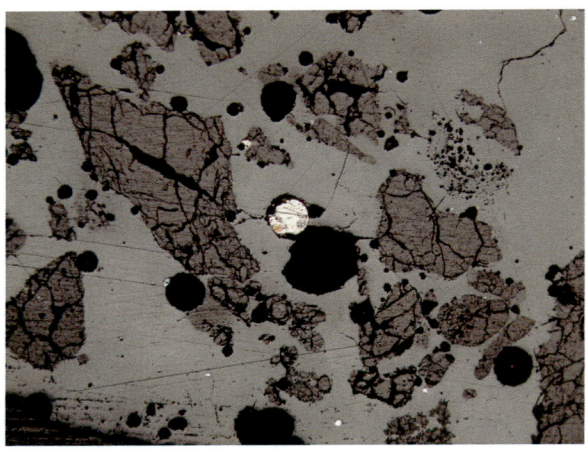

Low magnification image of lumpy a slag sample showing a polycrystalline aggregate of un-reacted quartz grains together with corroded metal prills in a glassy matrix. x 62.5.

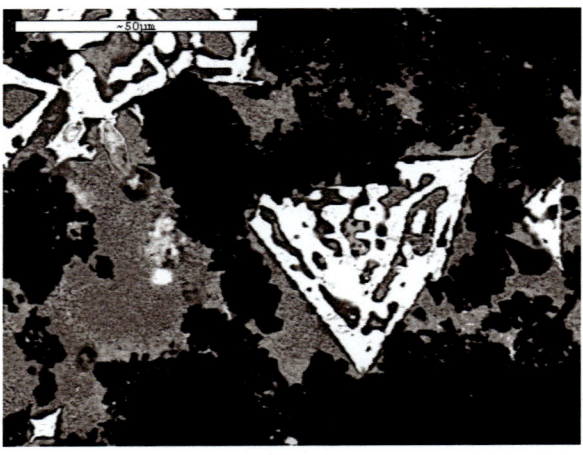

SEM micrograph of magnetite crystals in a large lump of decomposing cores.

Sample 1-2a: SEM analysis of magnetite crystals.

SEMQuant results. Listed at 3:41:16 PM on 8/10/04
Operator: blanca maldonado
Client: none
Job: blanca maldonado
Spectrum label: PSC-04-1/II-2A-clstr6/FeO-1

System resolution = 81 eV

Quantitative method: ZAF (2 iterations).
 Analysed elements combined with: O (Valency: -2)
 Method : Stoichiometry Normalised results.
 Nos. of ions calculation based on 32 anions per formula.

1 peak possibly omitted: 0.00 keV

Standards :
Na K   NAK 6/11/04
Mg K   MgO 01/12/93
Al K   alk 5/26/04
Si K   siwol 5/25/04
P K   GaP 29/11/93
S K   sk 12/13/01
K K   KK 6/11/04
Ca K   cak 5/25/04
Ti K   Ti 01/12/93
Mn K   mnk 12/14/01
Fe K   fek 5/19/04
Ni K   nil 5/19/04
Cu K   cul 5/19/04
Zn K   znk 5/18/04
Pb L   Pbl 11/11/98

Elmt	Spect. Type	Element %	Atomic %	Compound	Compound %	Nos. of ions
Na K	ED	0.83	1.22	Na2O	1.12	0.77
Mg K	ED	0.16	0.22	MgO	0.26	0.14
Al K	ED	1.74	2.20	Al2O3	3.30	1.38
Si K	ED	0.35	0.43	SiO2	0.76	0.27
P K	ED	0.02*	0.02*	P2O5	0.05*	0.01*
S K	ED	0.01*	0.01*	SO3	0.02*	0.01*

APPENDIX C: SLAG ANALYSES

Elmt	Spect. Type	Element %	Atomic %	Compound	Compound %
K K	ED	0.07	0.06	K2O	0.09
					0.04
Ca K	ED	0.02*	0.01*	CaO	0.02*
					0.01*
Ti K	ED	1.42	1.01	TiO2	2.37
					0.63
Mn K	ED	-0.07*	-0.04*	MnO	-0.09*
					-0.03*
Fe K	ED	72.39	44.03	FeO	93.12
					27.64
Ni K	ED	0.02*	0.01*	NiO	0.03*
					0.01*
Cu K	ED	0.00*	0.00*	CuO	-0.01*
					0.00*
Zn K	ED	0.02*	0.01*	ZnO	0.03*
					0.01*
Pb L	ED	-0.99*	-0.16*	PbO	-1.06*
					-0.10*
O		24.01	50.97		32.00
Total		100.00	100.00		100.00

Cation sum 30.78
* = <2 Sigma

Sample 1-2a: SEM analysis of magnetite crystals.

SEMQuant results. Listed at 3:43:14 PM on 8/10/04
Operator: blanca maldonado
Client: none
Job: blanca maldonado
Spectrum label: PSC-04-1/II-2A-clstr6/FeO-2

System resolution = 81 eV

Quantitative method: ZAF (2 iterations).
 Analysed elements combined with: O (Valency: -2)
 Method : Stoichiometry Normalised results.
 Nos. of ions calculation based on 32 anions per formula.

 1 peak possibly omitted: 0.00 keV

Standards :
  Na K   NAK 6/11/04
  Mg K   MgO 01/12/93
  Al K   alk 5/26/04
  Si K   siwol 5/25/04
  P K   GaP 29/11/93
  S K   sk 12/13/01
  K K   KK 6/11/04
  Ca K   cak 5/25/04
  Ti K   Ti 01/12/93
  Mn K   mnk 12/14/01
  Fe K   fek 5/19/04
  Ni K   nil 5/19/04
  Cu K   cul 5/19/04
  Zn K   znk 5/18/04
  Pb L   Pbl 11/11/98

Elmt	Spect. Type	Element %	Atomic %	Compound	Compound %
Na K	ED	0.78	1.16	Na2O	1.06
					0.73
Mg K	ED	0.14	0.20	MgO	0.23
					0.12
Al K	ED	1.77	2.23	Al2O3	3.34
					1.40
Si K	ED	0.23	0.28	SiO2	0.49
					0.17
P K	ED	0.00*	0.00*	P2O5	0.01*
					0.00*
S K	ED	0.00*	0.00*	SO3	0.00*
					0.00*
K K	ED	0.02*	0.02*	K2O	0.03*
					0.01*
Ca K	ED	0.02*	0.02*	CaO	0.03*
					0.01*
Ti K	ED	1.55	1.10	TiO2	2.59
					0.69
Mn K	ED	0.03*	0.02*	MnO	0.03*
					0.01*
Fe K	ED	72.28	44.05	FeO	92.99
					27.66
Ni K	ED	0.06*	0.04*	NiO	0.08*
					0.02*
Cu K	ED	0.15	0.08	CuO	0.19
					0.05
Zn K	ED	0.05*	0.03*	ZnO	0.06*
					0.02*
Pb L	ED	-1.06*	-0.17*	PbO	-1.14*
					-0.11*
O		23.95	50.95		32.00
Total		100.00	100.00		100.00

Cation sum 30.80
* = <2 Sigma

Sample 1-2a: SEM/EDS analysis of two magnetite crystals.

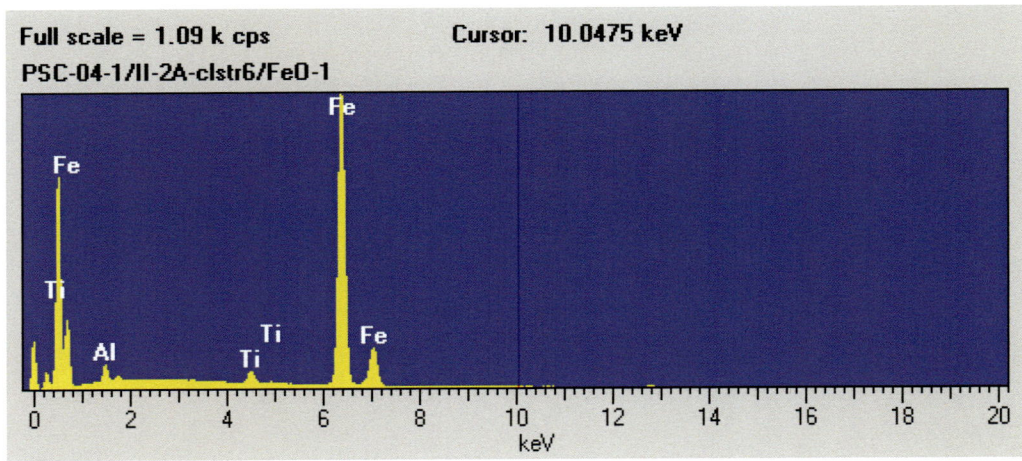

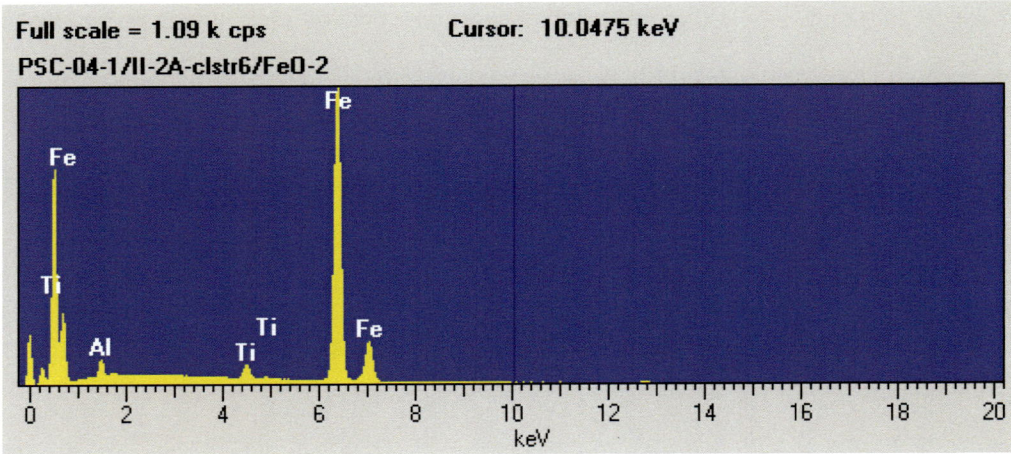

Sample 1-3a: Light Microscopy and SEM/EDS Analyses

Polished thin section.

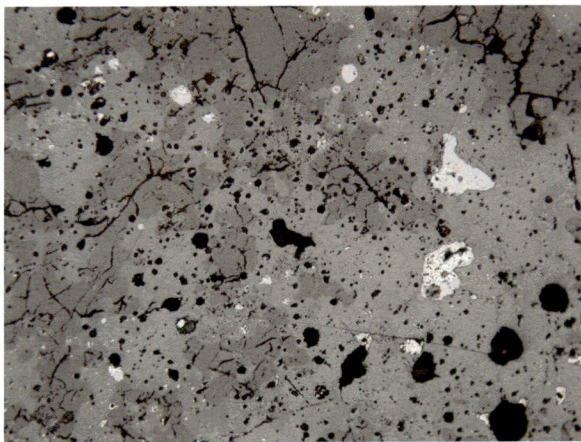

Micrograph showing iron sulfides in association with lumps of chalcopyrite and copper sulfides. x 62.5.

APPENDIX C: SLAG ANALYSES

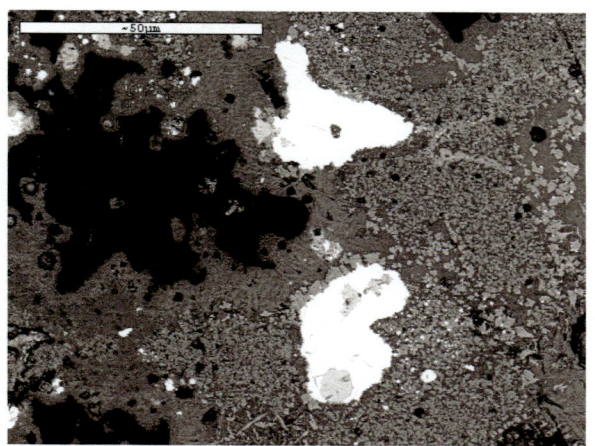

SEM micrograph showing iron sulfides in association with lumps of chalcopyrite and copper sulfides.

Sample 1-3a: SEM analysis of chalcopyrite lump.

SEMQuant results. Listed at 1:08:14 PM on 8/11/04
Operator: blanca maldonado
Client: none
Job: blanca maldonado
Spectrum label: PSC-04-1-1/II-3A-clstr2/CuS-1

System resolution = 83 eV

Quantitative method: ZAF (3 iterations).
Analysed all elements and normalised results.

4 peaks possibly omitted: 0.00, 0.26, 0.66, 0.92 keV

Standards:
S K	sk	12/13/01
Fe K	fek	5/19/04
Ni K	nil	5/19/04
Cu K	cul	5/19/04
Zn K	znk	5/18/04
As K	ask	12/13/01
Sn L	snl	5/19/04
Pb L	Pbl	11/11/98

Elmt	Spect. Type	Element %	Atomic %
S K	ED	29.74	44.84
Fe K	ED	16.18	14.01
Ni K	ED	0.00*	0.00*
Cu K	ED	54.23	41.27
Zn K	ED	0.10*	0.08*
As K	ED	-0.28*	-0.18*
Sn L	ED	-0.14*	-0.06*
Pb L	ED	0.17*	0.04*
Total		100.00	100.00

* = <2 Sigma

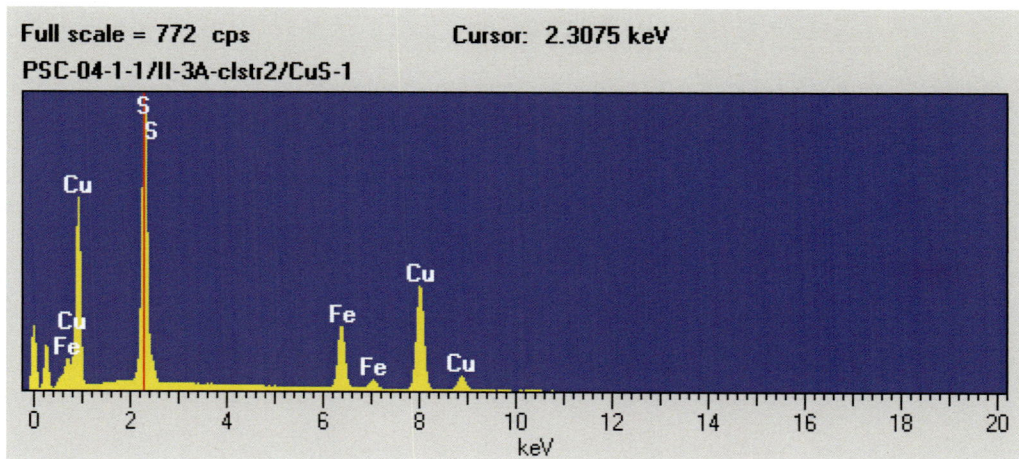

Sample 1-3a: SEM analysis of iron sulfides.

SEMQuant results. Listed at 3:57:03 PM on 8/10/04
Operator: blanca maldonado
Client: none
Job: blanca maldonado
Spectrum label: PSC-04-1/II-3A-prll/CuS-5

System resolution = 81 eV

Quantitative method: ZAF (2 iterations).
 Analysed all elements and normalised results.

 5 peaks possibly omitted: 0.00, 0.26, 0.68,
  1.44, 1.78 keV

Standards :
S   K   sk 12/13/01
Fe K   fek 5/19/04
Ni K   nil 5/19/04
Cu K   cul 5/19/04
Zn K   znk 5/18/04
As K   ask 12/13/01
Sn L   snl 5/19/04
Pb L   Pbl 11/11/98

Elmt	Spect. Type	Element %	Atomic %
S K	ED	38.59	51.31
Fe K	ED	62.42	47.64
Ni K	ED	0.07*	0.05*
Cu K	ED	1.24	0.83
Zn K	ED	0.42	0.27
As K	ED	1.25	0.71
Sn L	ED	0.06*	0.02*
Pb L	ED	-4.04*	-0.83*
Total		100.00	100.00

* = <2 Sigma

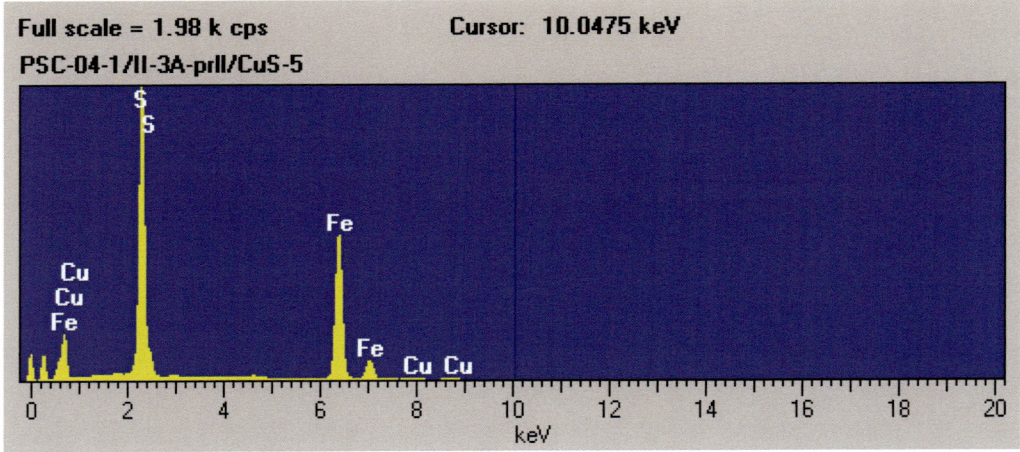

Sample 1-3a: Light Microscopy and SEM/EDS Analyses

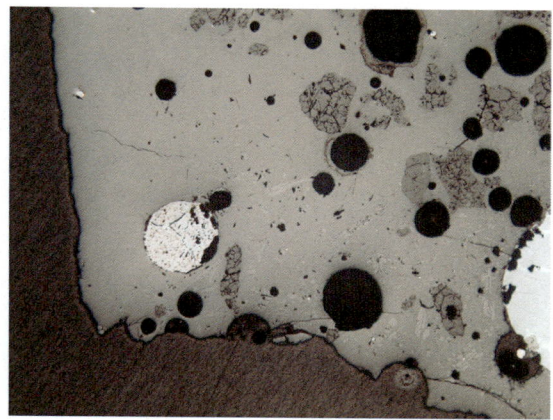

1) Micrograph showing a sulfide prill in a glassy matrix. x 62.

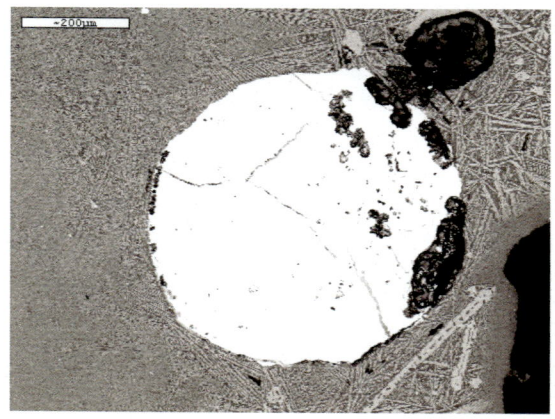

SEM image of sulfide prill a glassy matrix.

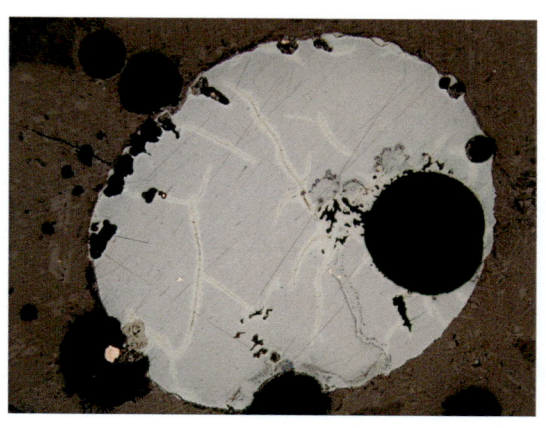

2) Micrograph showing a sulfide prill precipitating into metal. x 400.

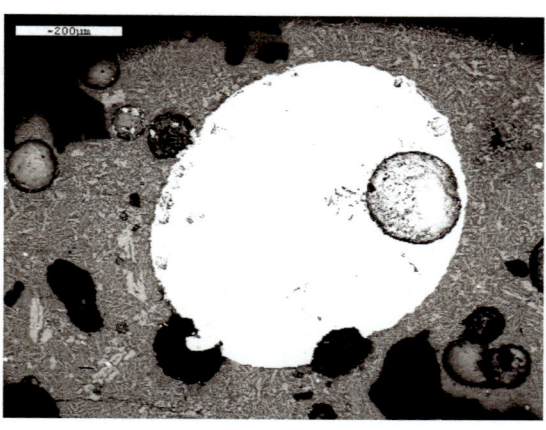

SEM image of precipitating prill.

3) Cluster of skeletal magnetite crystals. x 62

Cluster of skeletal magnetite crystals. x 125.

Sample 1-3a: SEM analysis of sulfide prill.

SEMQuant results. Listed at 3:52:47 PM on 8/10/04
Operator: blanca maldonado
Client: none
Job: blanca maldonado
Spectrum label: PSC-04-1/II-3A-prll/CuS-1

System resolution = 81 eV

Quantitative method: ZAF (3 iterations).
 Analysed all elements and normalised results.

 6 peaks possibly omitted: 0.00, 0.26, 0.66,
  0.92, 1.46, 1.80 keV

Standards :
S K	sk	12/13/01
Fe K	fek	5/19/04
Ni K	nil	5/19/04
Cu K	cul	5/19/04
Zn K	znk	5/18/04
As K	ask	12/13/01
Sn L	snl	5/19/04
Pb L	Pbl	11/11/98

Elmt	Spect. Type	Element %	Atomic %
S K	ED	27.33	41.98
Fe K	ED	15.46	13.63
Ni K	ED	-0.06*	-0.05*
Cu K	ED	57.58	44.63
Zn K	ED	0.22	0.17
As K	ED	-0.53*	-0.35*
Sn L	ED	-0.04*	-0.02*
Pb L	ED	0.05*	0.01*
Total		100.00	100.00

* = <2 Sigma

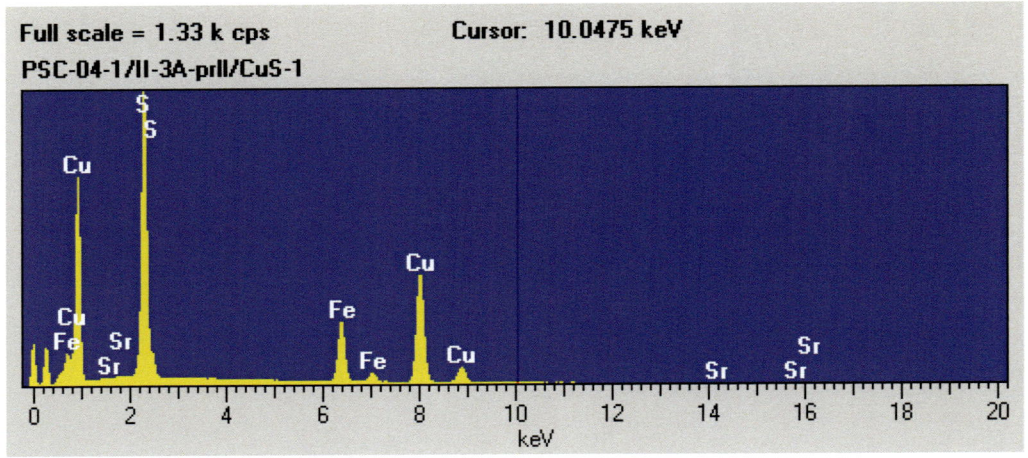

Appendix C: Slag analyses

Sample 1-3a: SEM analysis of precipitating prill.

SEMQuant results. Listed at 3:54:51 PM on 8/10/04
Operator: blanca maldonado
Client: none
Job: blanca maldonado
Spectrum label: PSC-04-1/II-3A-prll/CuS-2

System resolution = 81 eV

Quantitative method: ZAF (3 iterations).
Analysed all elements and normalised results.

6 peaks possibly omitted: 0.00, 0.26, 0.66,
0.92, 1.46, 1.80 keV

Standards :
S K	sk	12/13/01
Fe K	fek	5/19/04
Ni K	nil	5/19/04
Cu K	cul	5/19/04
Zn K	znk	5/18/04
As K	ask	12/13/01
Sn L	snl	5/19/04
Pb L	Pbl	11/11/98

Elmt	Spect. Type	Element %	Atomic %
S K	ED	24.93	39.12
Fe K	ED	11.88	10.70
Ni K	ED	-0.07*	-0.06*
Cu K	ED	63.80	50.51
Zn K	ED	0.19	0.15
As K	ED	-0.54*	-0.36*
Sn L	ED	-0.03*	-0.01*
Pb L	ED	-0.16*	-0.04*
Total		100.00	100.00

* = <2 Sigma

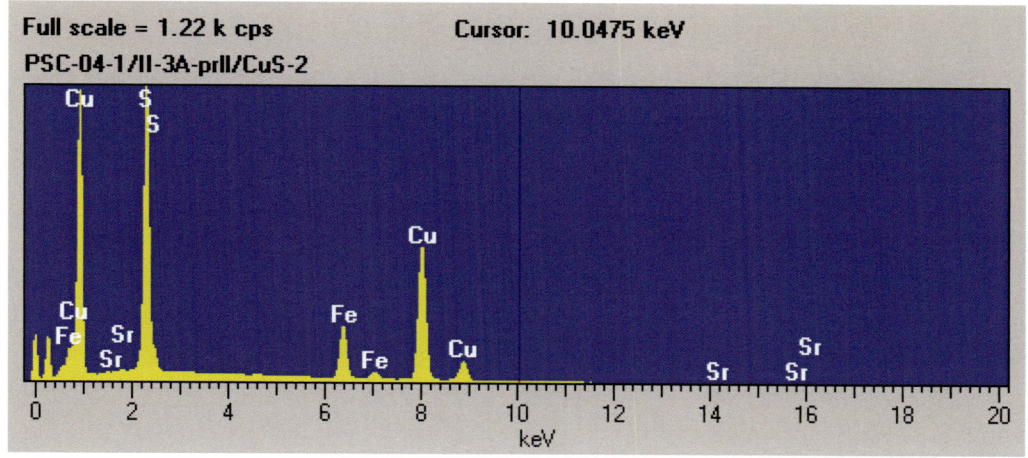

Sample 1-3b: Light Microscopy and SEM/EDS Analyses

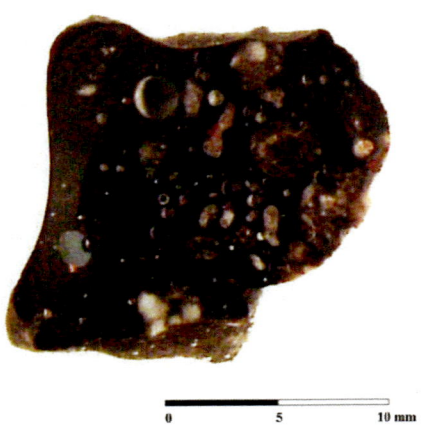

The above precipitating prill x 400.

Polished thin section

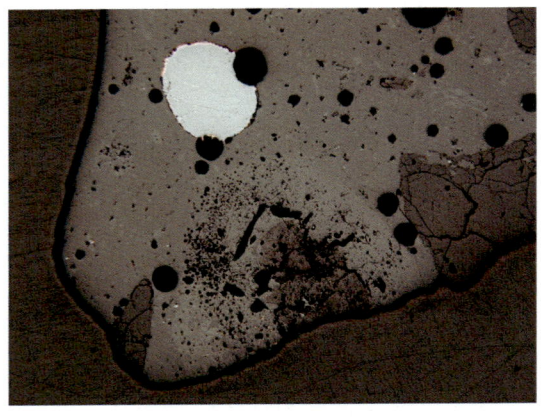

2) Micrograph showing a sulfide prill and unreacted materials in a glassy matrix. x 62.

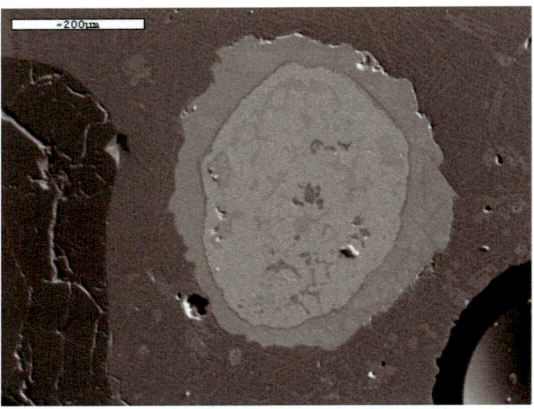

SEM image of above copper and iron sulfide.

1) Micrograph showing a sulfide prill x 62.

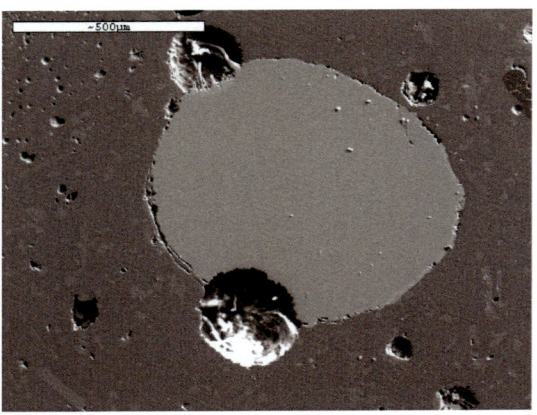

SEM image of precipitating prill.

Sample 1-3b: SEM analysis of copper and iron sulfide.

SEMQuant results. Listed at 3:52:51 PM on 8/12/04
Operator: blanca maldonado
Client: none
Job: blanca maldonado
Spectrum label: PSC-04-1/II-3B-prll3-4

System resolution = 82 eV

Quantitative method: ZAF (3 iterations).
Analysed all elements and normalised results.

5 peaks possibly omitted: 0.02, 0.28, 0.66,
0.92, 1.78 keV

Standards :
S K	sk	12/13/01
Fe K	fek	5/19/04
Ni K	nil	5/19/04
Cu K	cul	5/19/04
Zn K	znk	5/18/04
As K	ask	12/13/01
Sn L	snl	5/19/04
Pb L	Pbl	11/11/98

Elmt	Spect. Type	Element %	Atomic %
S K	ED	29.01	43.99
Fe K	ED	14.79	12.88
Ni K	ED	0.00*	0.00*
Cu K	ED	56.49	43.23
Zn K	ED	-0.06*	-0.04*
As K	ED	0.03*	0.02*
Sn L	ED	-0.08*	-0.03*
Pb L	ED	-0.19*	-0.04*
Total		100.00	100.00

* = <2 Sigma

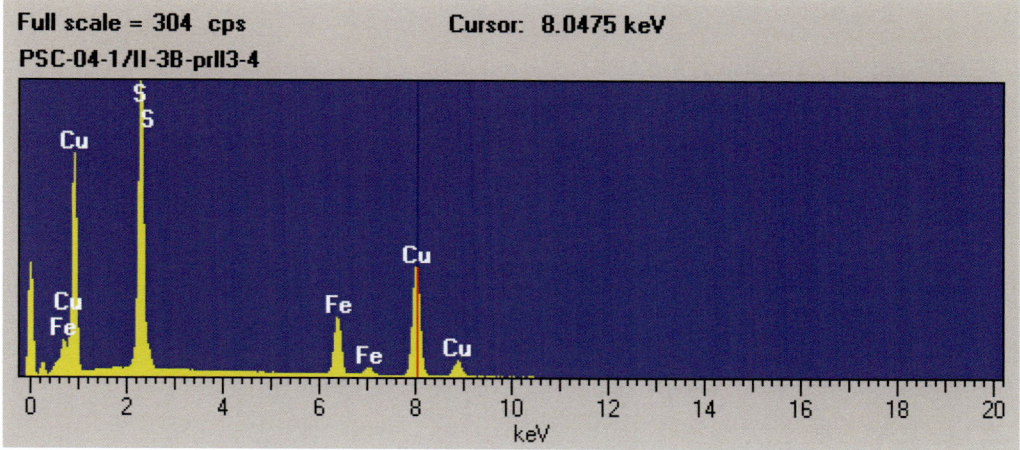

Sample 1-3b: SEM analysis of precipitating prill.

SEMQuant results. Listed at 1:39:05 PM on 8/12/04
Operator: blanca maldonado
Client: none
Job: blanca maldonado
Spectrum label: PSC-04-1/II-3B-prll/CuS-1

System resolution = 82 eV

Quantitative method: ZAF (3 iterations).
Analysed all elements and normalised results.

4 peaks possibly omitted: 0.02, 0.28, 0.66, 0.92 keV

Standards :

S K	sk	12/13/01
Fe K	fek	5/19/04
Ni K	nil	5/19/04
Cu K	cul	5/19/04
Zn K	znk	5/18/04
As K	ask	12/13/01
Sn L	snl	5/19/04
Pb L	Pbl	11/11/98

Elmt	Spect. Type	Element %	Atomic %
S K	ED	26.64	41.27
Fe K	ED	13.73	12.21
Ni K	ED	-0.01*	-0.01*
Cu K	ED	59.60	46.59
Zn K	ED	0.13*	0.10*
As K	ED	-0.29*	-0.19*
Sn L	ED	-0.06*	-0.03*
Pb L	ED	0.27*	0.06*
Total		100.00	100.00

* = <2 Sigma

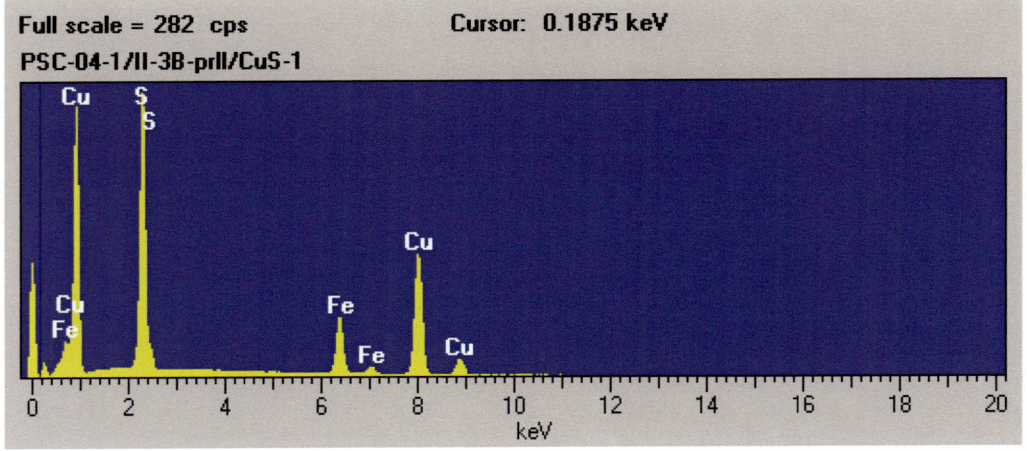

Sample 2-1a: Light Microscopy and SEM/EDS Analyses

Polished thin section.

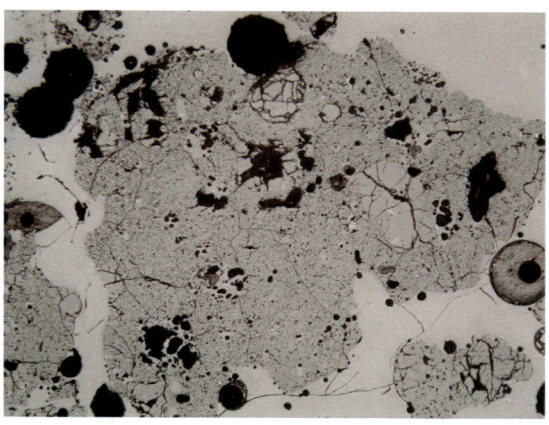

Micrograph showing un-reacted material in glassy matrix.

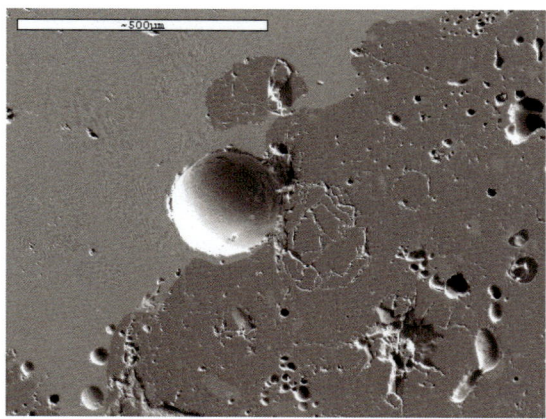

SEM micrograph showing un-reacted material in glassy matrix.

Sample 2-1a: SEM analysis of glassy matrix.

```
SEMQuant results. Listed at 10:10:32
AM on 8/17/04
Operator: blanca maldonado
Client: none
Job: blanca maldonado
Spectrum label: PSC-04-2/I-1A-mtrx-1

System resolution =   83 eV

Quantitative method: ZAF (3
iterations).
 Analysed elements combined with: O
(Valency: -2)
 Method : Stoichiometry Normalised
results.
 Nos. of ions calculation based on 32
anions per formula.

1 peak possibly omitted: 0.00 keV

Standards :
 Na K      NAK 6/11/04
 Mg K      MgO 01/12/93
 Al K      alk 5/26/04

 Si K      siwol 5/25/04
 P  K      GaP 29/11/93
 S  K      sk 12/13/01
 K  K      KK 6/11/04
 Ca K      cak 5/25/04
 Ti K      Ti 01/12/93
 Mn K      mnk 12/14/01
 Fe K      fek 5/19/04
 Ni K      nil 5/19/04
 Cu K      cul 5/19/04
 Zn K      znk 5/18/04
 Pb L      Pbl 11/11/98
```

Elmt	Spect. Type	Element %	Atomic %	Compound	Nos. of ions
Na K	ED	0.68	0.74	Na2O 0.91	0.40
Mg K	ED	1.07	1.11	MgO 1.78	0.60
Al K	ED	5.89	5.48	Al2O3 11.13	2.95
Si K	ED	17.81	15.92	SiO2 38.09	8.57
P K	ED	0.01*	0.01*	P2O5 0.02*	0.00*
S K	ED	0.55	0.43	SO3 1.38	0.23
K K	ED	1.20	0.77	K2O	

Sample 2-1a: SEM analysis of unreacted material.

SEMQuant results. Listed at 11:46:36 AM on 8/17/04
Operator: blanca maldonado
Client: none
Job: blanca maldonado
Spectrum label: PSC-04-2/I-1A-clstr1-lmp

System resolution = 82 eV

Quantitative method: ZAF (2 iterations).
 Analysed elements combined with: O (Valency: -2)
 Method : Stoichiometry Normalised results.
 Nos. of ions calculation based on 32 anions per formula.

1 peak possibly omitted: 0.00 keV

Standards :
Na K   NAK 6/11/04
Mg K   MgO 01/12/93
Al K   alk 5/26/04
Si K   siwol 5/25/04
P K   GaP 29/11/93
S K   sk 12/13/01
K K   KK 6/11/04
Ca K   cak 5/25/04
Ti K   Ti 01/12/93
Mn K   mnk 12/14/01
Fe K   fek 5/19/04
Ni K   nil 5/19/04
Cu K   cul 5/19/04
Zn K   znk 5/18/04
Sn L   snl 5/19/04
Pb L   Pbl 11/11/98

Elmt	Spect. Type	Element %	Atomic %	Compound	Compound %	Nos. of ions
Na K	ED	0.06	0.05	Na2O	0.08	0.02
Mg K	ED	-0.07*	-0.06*	MgO	-0.11*	-0.03*
Al K	ED	-0.15*	-0.11*	Al2O3	-0.28*	-0.05*
Si K	ED	46.64	33.30	SiO2	99.77	15.96
P K	ED	-0.08*	-0.05*	P2O5	-0.18*	-0.02*
S K	ED	0.22	0.14	SO3	0.56	0.07
K K	ED			K2O	1.44	0.41
Ca K	ED	2.36	1.48	CaO	3.30	0.80
Ti K	ED	0.12	0.06	TiO2	0.20	0.03
Mn K	ED	0.11	0.05	MnO	0.14	0.03
Fe K	ED	32.14	14.45	FeO	41.34	7.78
Ni K	ED	-0.03*	-0.01*	NiO	-0.04*	-0.01*
Cu K	ED	0.22	0.09	CuO	0.28	0.05
Zn K	ED	0.02*	0.01*	ZnO	0.03*	0.00*
Pb L	ED	0.00*	0.00*	PbO	0.00*	0.00*
O		37.86	59.42			32.00
Total		100.00	100.00		100.00	

Cation sum 21.85
* = <2 Sigma

```
K   K    ED   -0.14*   -0.07*   K2O      0.08*         0.01*
-0.17*   -0.03*                 Sn  L    ED     0.52      0.09    SnO2
Ca  K    ED   -0.14*   -0.07*   CaO      0.67          0.04
-0.20*   -0.03*                 Pb  L    ED    -0.20*   -0.02*    PbO
Ti  K    ED    0.01*    0.00*   TiO2    -0.22*        -0.01*
 0.01*    0.00*                 O              53.27    66.78
Mn  K    ED    0.00*    0.00*   MnO     32.00
 0.00*    0.00*                 Total         100.00   100.00
Fe  K    ED    0.02*    0.01*   FeO    100.00
 0.02*    0.00*
Ni  K    ED   -0.01*    0.00*   NiO     Cation sum 15.92
-0.01*    0.00*                 * = <2 Sigma
Cu  K    ED   -0.01*    0.00*   CuO
-0.01*    0.00*
Zn  K    ED    0.06*    0.02*   ZnO
```

Sample 2-1a: SEM/EDS analysis

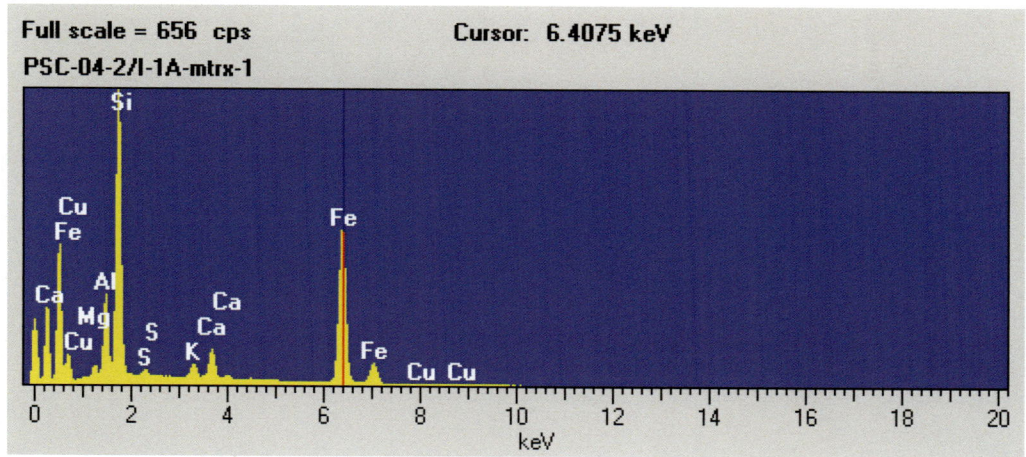

SEM analysis of glassy matrix.

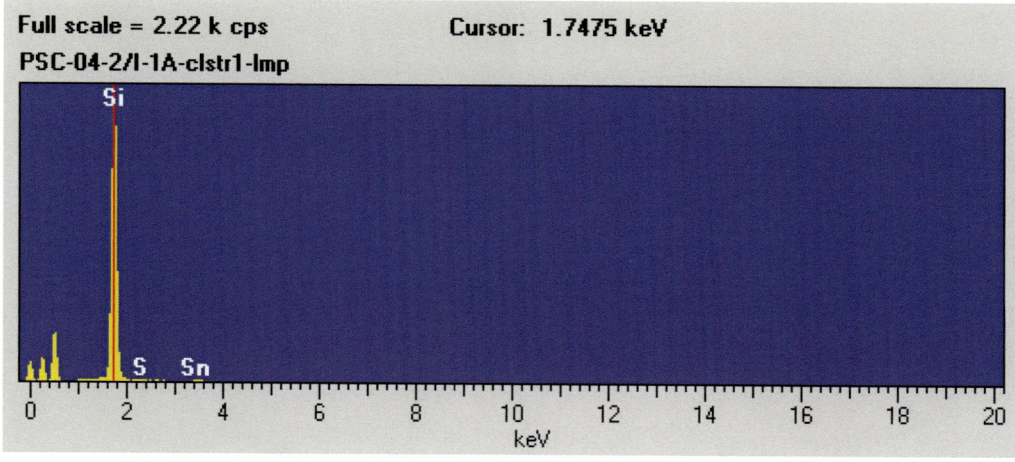

SEM analysis of un-reacted material.

Tarascan Copper Metallurgy

Polished thin sections of samples not analyzed by SEM/EDS

5-1a

7-1

7-1b

3-1a

3-1c

J-sa

J-sb

J-sc

Appendix C: Slag Analyses

XRF Bulk Composition of Pressed Pellets

XRF AVERAGE VALUES WITH ORIGINAL TOTALS GIVEN

CONTEXT	SLAG	SiO_2 %	Al_2O_3 %	FeO %	TiO_2 %	MnO %	CaO %	MgO %	Na_2O %	K_2O %	P_2O_5 %	SO_3 %	CuO ppm	PbO ppm	ZnO ppm	TOTAL
1-1a	p	31.2	8.61	42.3	0.28	0.11	0.90	1.16	0.52	1.16	0.87	0.12	30440	16	493	90.3
1-1b	p	24.9	4.14	68.2	0.08	0.12	1.64	0.99	0.48	0.77	0.02	2.35	22940	0	401	106
1-2b	p	32.4	10.9	46.9	0.35	0.13	1.46	2.06	0.56	0.82	0.11	0.27	12913	225	1061	97.2
1-4a	p	35.0	9.65	43.7	0.29	0.09	1.72	2.30	0.61	1.11	0.07	0.26	14787	0	641	96.2
2-1b	p	40.0	3.44	46.2	0.04	0.13	4.29	0.61	0.56	0.50	0.01	0.24	14387	1544	1099	97.7
1-2a	l	56.4	7.78	30.5	0.27	0.07	0.89	1.29	0.41	1.47	0.05	0.48	15720	0	147	101
1-3a	l	55.9	7.59	30.3	0.23	0.06	1.71	1.30	0.44	1.34	0.05	0.42	12317	0	154	100
1-3b	l	41.2	7.56	42.8	0.25	0.09	0.93	0.96	0.56	0.96	0.10	0.46	11147	381	1359	97.0
2-1a	l	54.4	11.1	27.5	0.23	0.08	2.50	1.44	0.35	1.95	0.04	0.71	7748	0	267	101
3-1a	l	57.0	5.04	32.7	0.15	0.06	0.59	0.52	0.21	0.84	0.03	0.35	18673	399	620	99.4
3-1b	l	41.6	7.12	44.0	0.24	0.07	2.23	1.46	0.57	1.06	0.07	0.71	32797	147	2809	102
3-1c	l	61.7	3.96	32.6	0.12	0.04	0.48	0.03	0.08	0.87	0.06	0.54	9271	303	2543	101
ECRM 681-1		17.7	12.3	41.8	0.46	0.27	3.65	1.28	0.39	0.55	1.75	0.14	28	0	444	80.6
BCS CRM 301/1		7.61	4.27	31.7	0.16	1.12	20.7	1.43	0.03	0.33	0.62	0.05	33	17	185	68.1
BCS CRM 381/1		8.78	0.00	17.00	0.31	2.85	48.5	0.0	0.74	0.05	12.7	0.00	21	0	19	92.1

Phase diagrams

The determination of the mineralogical and chemical composition of copper smelting slags is of interest to archaeometallurgists because it can help determine the most important parameters of the smelting process, namely temperature, oxygen partial pressure, and cooling rate. These parameters can help to reconstruct ancient smelting processes and provide information on the design and operation of ancient copper smelting furnaces. The major components in preindustrial slags are typically FeO, CaO, and SiO_2. In addition, slag may contain variable amounts of MgO, MnO, Al_2O_3, P_2O_5 and other compounds, as well as minute inclusions of metal. The relationship between chemical composition and compound formation during and after the solidification of slags is often investigated through phase diagrams (Bachmann 1982).

Basic Definitions:

1. A *phase* is a homogeneous portion of a system that has uniform physical and chemical characteristics. Two distinct phases in a system have distinct physical or chemical characteristics (e.g. water and ice) and are separated from each other by definite phase boundaries. A phase may contain one or more components.
2. A *component* is a substance (chemical compound or element) that interacts with other compounds forming phases (Fe and C in carbon steel, H_2O and NaCl in salted water).
3. A *system* is a specific body (volume) that cannot exchange matter with surrounding.
4. An *alloy* is a phase comprising of one or more components. A binary alloy contains two components, a ternary alloy – three, etc. A binary alloy contains two components, a ternary alloy three, etc.
5. The *concentration* of component causing an alloy to form more than one phase is called *solubility limit*. If both phases remain solid, then the product is called multiphase alloy (composite). For example: Brass (Zn:Cu) is an alloy, Steel is a composite.

Definition of a phase diagram:

1. *Phase equilibrium* is state of the system, at which all its parameters remain constant in time.
2. A system for which the rate of achieving equilibrium is very low is called *metastable*.
3. A *phase diagram* (often termed an equilibrium or constitutional diagram) graphical representation of the combinations of temperature, pressure, composition, or other variables for which specific phases exist at equilibrium.
4. Phase separation is the transformation of a homogenous system into two (or more) phases.
5. The point at which liquid solidifies in two phases is called *eutectic* point.

Note: The above definitions have been extracted from the General Chemistry Glossary (Copyright © 1997-2005 by Fred Senese): *http://antoine.frostburg.edu/chem/senese/101/glossary.shtml*

Chemical and Phase Composition of Archaeological Slags

For a given temperature and composition we can use phase diagrams to determine:

1. The phases that are present
2. Compositions of the phases
3. The relative fractions of the phases

Phase transformations leading to formation of two solid phases are of great importance in archaeometallurgy. Eutectic structure formation (liquid to solid transition) is the most relevant type of phase transformation for slag analysis. The use of ternary phase diagrams is common method to represent analytically determined compositions of slags, often presented as oxide percentages (e.g. SiO_2, FeO, Al_2O_3, etc.). The following explanation has been adapted from Hans-Gert Bachmann's The Identification of Slags from Archaeological Sites (1982).

A three-component system can be characterized by means of triangular coordinates for the composition and uniform temperature intervals by isotherms in the same way as elevation contours are drawn on topographical maps. Figure C.1 illustrates how the composition of a ternary system is represented by triangular coordinates. Each side of an equilateral triangle is divided into ten parts (one part equals ten weight percent), each division being intersected by lines parallel to each of the other two sides. The point C at the apex of the triangle is composed wholly of component C (equaling 100 weight percent). Any point on the base line A-B is composed entirely of components A and B with none of C. The relative distance of point X from each of the three apices may be expressed as a percentage and thus represents the percentage composition of a ternary mixture in terms of components A, B, and C. Point X, for example, has a composition of 40 percent A, 20 percent B, and 40 percent C.

Phase diagrams are graphic simulations and therefore have limitations. Apart from the fact that they represent systems in equilibrium, only the major oxide components of slags can be included in ternary diagrams. In the case of the samples from Itziparátzico for instance, only the most representative components (SiO_2, FeO, Al_2O_3; see Figure 4.2) were considered. Nevertheless, important information concerning thermodynamics, crystallization and sequences, activity of components in compound formation, etc. can be extracted from these graphic devices. The treatment of metallurgical process, both modern and ancient, is incomplete without the implications of phase relations and properties provided by phase diagrams. The substantial progress in archaeometallurgical studies seen in the last few decades has been made possible, in part, by the interpretive possibilities offered by this branch of physical chemistry (Bachmann 1982: 11).

The crystallization path (i.e., the composition of the liquid as a function of temperature as the melt cooled) in a simple theoretical ternary system may illustrate the deductions that can be drawn from ternary phase diagrams. Knowledge of the crystallization curve or the melting curve (the reverse of crystallization curve) for any particular melt is extremely valuable in the study of phase formation during the solidification of slags. In Figure C.2, point m is the 'ternary eutectic', i.e. the point at which all the crystallization curves of this system terminate. If a slag melt of composition a is chosen and allowed to cool, the system remains liquid until the 'liquidus' (the more or less curved plane separating the liquid from the solid) is reached. At this temperature the solid A begins to crystallize. The course of the crystallization from this point to the boundary m-k follows a straight line drawn through A and a. Due to crystallization of component A, the melt alters its composition. As the liquid melt changes in composition from a to b, more and more A crystallizes. At b, a second mineral phase appears and the crystallization curve follows boundary k-m with phases A and C crystallizing together. At point m, the temperature remains constant with slag minerals A, B, and C crystallizing simultaneously until the remainder of the molten slag has solidified. The final product will be a mixture of large crystals of A and C and small crystals ('eutectic mixture') of A, B, and C.

The composition of the solid slag minerals crystallizing at a certain moment during the process of cooling along b-m is given at the point where the tangent to the crystallization curve intersects that side of the composition triangle representing the two coexisting solid phases. For example, at b it is indicated by the intersection of the tangent to the curve m-b-k at the point b with the line A-C at point b'. The ratio of A to C is given by the ratio b'-C/b'-A. The mean composition of the two solid phases that have crystallized between points b and m is represented by the intersection of a line drawn through m and b and the side of the composition triangle at b'. In this case it is a mixture of A and C in the proportion b'-C/b'-A. The mean composition of the total solid slag that separates out between a and m (before B begins to solidify) is determined by drawing a line through m and a to the side of the composition triangle at a'. During eutectic crystallization at m, the composition of the total solids, i.e. all the portions of the slag having crystallized during the continuous fall in temperature, changes from a' to a, reaching the latter point as the last drop of the melt disappears.

APPENDIX C: SLAG ANALYSES

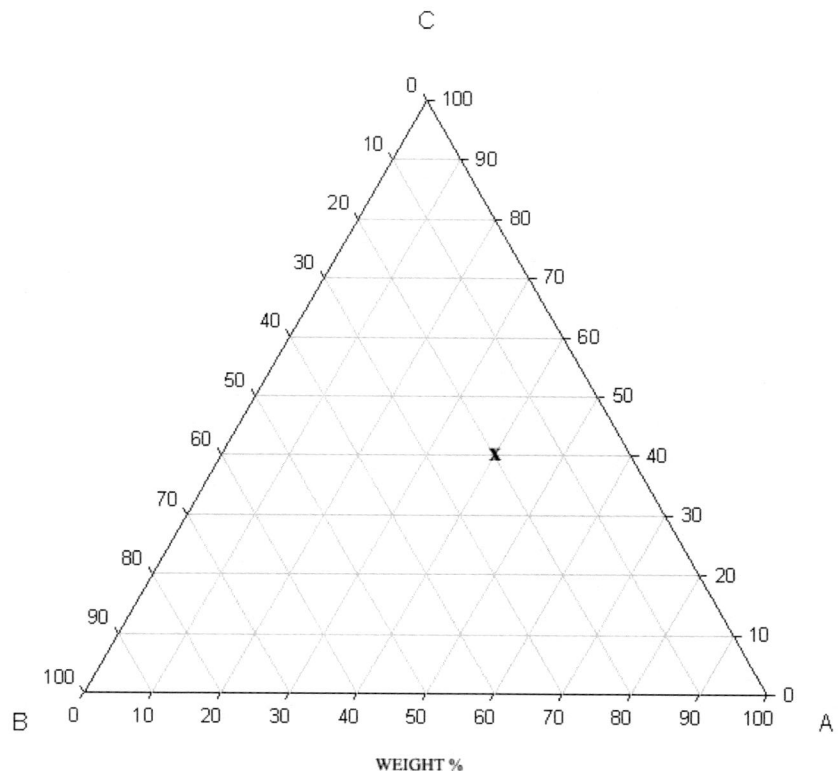

Figure C.1 Hypothetical ternary system A-B-C. Point X is shown to have a composition of 40 % A, 20 % B, and 40 % C
(After Bachmann 1982: Fig. 1).

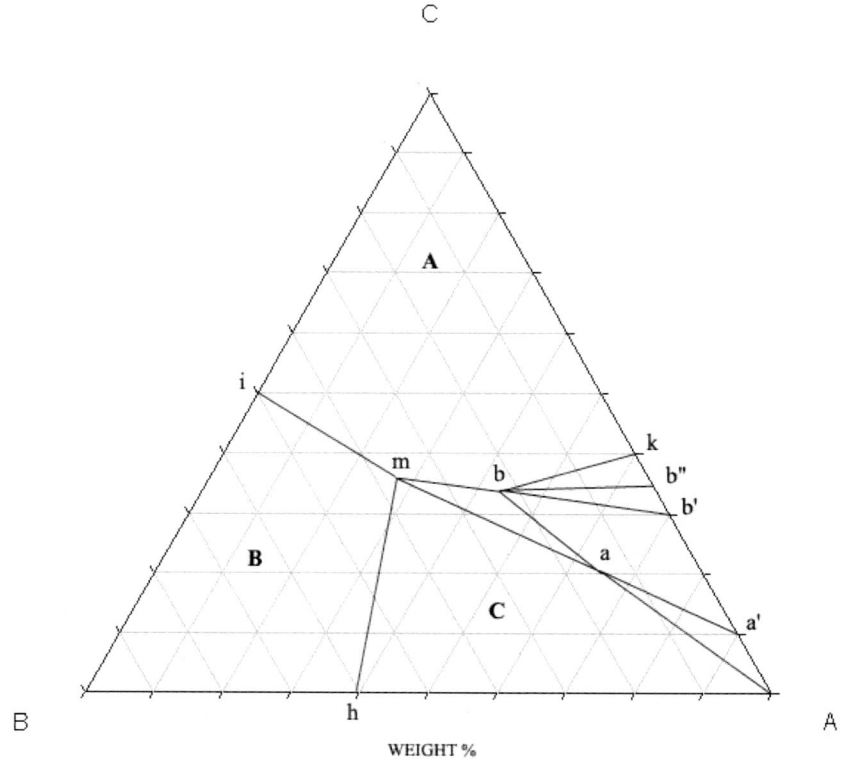

Figure C.2 Crystallization path in hypothetical ternary system A-B-C. Point m represents the ternary eutectic
(After Bachmann 1982: Fig. 3).

Appendix D

The rise of the theory of technology

This appendix is an extension of Chapter 2, outlining previous theoretical and field studies in order to clarify the various approaches that have been taken to the problem of technology and culture. The effect of implicit or explicit theoretical approaches, including evolutionary theories underlying much of nineteenth and twentieth century anthropological thought, on field and laboratory studies of early or non-industrialized technologies is considered.

V. Gordon Childe: Technology and Social Evolution

Childe's thinking on the significance of technology represented a major advance over the level of previous speculation and the depth of his analysis is reflected in the lifelong evolution of his views. As noted, Childe came eventually to modify many of his earlier views in response to a reevaluation of the archaeological data bearing on technology and culture. For the sake of clarity in the following discussion, a number of points argued at various times by Childe will be presented and discussed without any consideration of the later modifications in these views. A more complete examination of Childe's theories may be found in several recent studies (McNairn 1980; Trigger 1980; Green 1981; Manzanilla 1987; Daniel and Chippindale 1989; Harris 1994).

Technology, in Childe's view, was important primarily through its role in the creation of a surplus. Technological innovation allowed increases in the efficiency of resource use and production and the consequent possibility of production beyond the level of subsistence. Surpluses arose from this increased productive efficiency. More efficient technologies, resulting from technological innovation, allowed the accumulation of a material surplus. The evolutionary potential of technology lay in the possibilities for the use of this surplus. One possibility was the use of surplus wealth to the advantage of a segment of the population. The manipulation of surplus wealth by some group or individuals could lead to the formation and reinforcement of an elite with a privileged social, economic and political position. Power would tend to accumulate in the hands of those controlling the surplus. The trend to centralized power thus derived, according to Childe, from the creation of a surplus through technological means. The creation of a surplus also permitted economic specialization. A large part of the population was freed from subsistence tasks and both allowed and encouraged (through investment by the elite) to pursue a range of non-subsistence occupations. Technological innovation itself fostered specialization since the discovery of new techniques and materials allowed new occupations to develop. Cultural complexity increased as new economic roles and new opportunities for the use of surplus wealth were created. The key to the process, in Childe's view, was the potential inherent in technological change.

While continuing to emphasize the crucial links among technology, resource control, and property relations in cultural evolution, Childe modified his position several times in his writings. The creation of a surplus and its use remained central in Childe's thinking, but he early recognized (e.g. Childe 1944: 21-24) that technological innovation by itself does not result in the formation of a surplus. Technologies are social phenomena and surplus formation and use occur in the context of social relations (Childe 1954). The unilinealism of Morgan was considerably altered in Childe's later writings, with the recognition that the social relations, from which technologies emerged, were complex and not deterministically linked to technology. Childe's debt to Marx is apparent (see McNairn 1980: 79-91, 150-167; Trigger 1980: 91-135). The occasional ambiguity in Childe's ideas on the primacy of technology and the changes in these views reflect, in part, a corresponding confusion in the writings of Karl Marx and Friedrich Engels on the subject. Like Childe, Marx seems frequently to change his position to the point of contradiction on the causal significance of technology (e.g. Bober 1955: 3-28).

The Marxist Component in Childe's Theory of Technology

Marx distinguishes the means of production, or productive forces (technology in a broad sense: technique materials, instruments, facilities), from the mode of production, or relations of production, which arise in response to the necessities by which humans must exploit their environment to live. Although at various points Marx seems to assign primary evolutionary significance to the means of production, it is clear that in general, Marx and Engels believed that changes in the mode of production underlie social change (Marx 1904). Even the definition of the productive forces is expanded to include labor and production systems. The mode of production is thus not defined simply by technique but by labor and the products of labor as well (Marx 1955: 197-206).

Far from implying that simple technological innovation and change determine economic and social relations, as some followers of Marx have claimed (see Bober 1955: 11-12), the general thrust of the Marxist argument is that the mode of production (comprising the relations between humans and materials engendered by production) is responsible for the forms of society and that changes in these relations determine changes in social relations. While acknowledging the importance of technique and technological innovation, Marx and Engels clearly recognize that innovation and the adoption of innovations occur in the context of existing relations of production and the corresponding social, economic, and political relations. This position is effectively summarized by Marx (1904: 11-12):

> In the social production which men carry on they enter into definite relations that are indispensable and independent of their will; these relations of production correspond to a definite stage of development of their material powers of production. The sum total of these relations of production constitutes the economic structure of society - the real foundation, on which rise legal and political superstructures and to which correspond definite forms of social consciousness. The mode of production in material life determines the general character of the social, political, and spiritual processes of life.

Childe eventually broadened his concept of technology to include both the means and the mode of production and gave increasing importance to the mode of production and to the corresponding social relations in cultural evolution. Changes in the mode of production, according to this modified view, might produce changes in property relations and influence the emergence of class differentiation (Childe 1944: 21-24). Although surplus accumulation continued to play a prominent role in Childe's thinking, he came to deny that surpluses arose solely through the effects of technological innovation. In his later writings (e.g. 1950, 1951a, 1951b, 1954), Childe stressed that technologies are social products reflecting to some degree the existing social relations.

Technology and Craft Production

Childe noted that control of technology often was used as a political tool. Departing from this observation, he developed one of the first efforts to relate a given culture's technological development to its changing political and economic organization (Patterson 2005). Childe's basic premise concerning the development of craft specialization, and in particular that of metallurgy focused on the Bronze Age (Childe 1930, 1942). Metallurgy, according to Childe, constituted a specialization significantly distinct from the stone tool manufacture of the Neolithic. Bronze production was a complicated process and required full-time engagement by certain members of the community, who in consequence could no longer participate directly in agricultural production. The adoption of metal technology also required well developed trade networks.

Because of its complexity, this process could not have been independently invented in the Near East and Europe (Childe 1930: 23-24). The specialized knowledge required for metal extraction (i.e. smelting) was on the other hand predisposed for easy control by priest-kings or temple administrators. Childe believed that bronze artifacts may have originated as luxury items, but rapidly became indispensable. The alluvial plains of Mesopotamia were a resource-poor region, and stone tin and copper had to be imported. Under these conditions the more durable bronze items presented a cost-effective advantage over lithic tools (Childe 1930: 17). In Europe bronze also became a necessity. Bronze tools were used to clear forests and, as competition over land increased, bronze weapons provided an important advantage over lithic weaponry (Childe 1930: 192; McNairn 1980: 34; Trigger 1980: 70; Wailes 1996).

The shift from stone to metal technology ultimately undermined the economic self-sufficiency that characterized the Neolithic period (Childe 1951: 97, 1954: 46). People became dependent upon specialists who where probably not members of their immediate kinship network. Furthermore, given the highly localized deposits of raw material needed for bronze production, communities became dependent upon trade. This transition would therefore have involved the creation of a series of task-specific occupations and an extensive exchange network (Childe 1930: 8-11; Trigger 1980: 69). While Childe viewed bronze-working as crucial throughout the Old World, its role in social evolution was not constant from place to place. He believed for instance, that Near Eastern metallurgists operated as 'attached' specialists (e.g. Brumfiel and Earle 1987), their patrons consisting of the priest-kings and other temple elite. Surplus extraction in this context created a class society, separating individuals with 'theoretical' knowledge (kings, priests) from those with 'practical' and 'applied' abilities (producers, laborers) McNairn 1980: 28). This situation eventually led to technological stagnation, since those in power had access to a labor surplus and did not need to promote 'labor saving inventions' (Childe 1951a: 261). The result was a system that remained marginal to the developments of the Urban Revolution (Childe 1951a: 258).

Bronze smiths in Europe, in contrast, were rarely tied to a single patron (Childe 1958). The European subsistence economy, according to Childe, would have made it difficult to amass sufficient surplus to support attached specialists. Instead, European metallurgists were independent, itinerant producers, peddling their goods from place to place and often competing for consumers. This mobility and competition would also foster an increased bronze production, leading to what Childe (1958: 170) described as an industry of 'inventiveness and ingenuity'. Childe argued that the Bronze Age in Europe was characterized by a changing social order in which craftspersons were forced outside of the prevailing kinship system, since itinerant merchants would have no kin immediately available in times of need (Childe 1946: 25, cited in McNairn 1980:86; Wailes 1996). Thus, unlike the 'stagnant Oriental' situation, social and economic development in Europe followed a course of progress which subsequently foreshadowed the 'peculiarities of European polity in antiquity, the Middle Ages and Modern Times' (Childe 1958: 172).

Problems in Childe's Theory of Technology

There are several problems with Chide's views that stem largely from Childe's lack of familiarity with the archaeological data of the New World and the ethnographic literature. While much of the ethnographic information was unavailable to Childe, it seems that he deliberately ignored archaeological data from outside the nuclear area of Old World civilizations, evidence which may have forced a substantial revision of his theories (Harris 1968: 682-683; Trigger 1980: 124-128; Green 1981: 86-87). The major problem is that of the deterministic relationship between technology, surplus production, control of resource exploitation, and economic specialization. A more comprehensive view of the ethnographic and archaeological data has lead recent scholars to question some of Childe's theories. Three questions are particularly suggestive.

First, are technological innovations necessarily adopted? Ethnographic and historical information clearly indicate that innovations are not adopted if the perceived effect of their adoption is to increase per capita expenditure of energy (Pearson 1957; Boserup 1965; Lee and DeVore 1968). The possibility of increasing systemic efficiency (lowering the ratio of overall energy invested to energy return) is either not perceived or is not sufficient to overcome the reluctance of individuals to increase their work load. Only when a pressure (e.g. population pressure, environmental degradation) to increase systemic efficiency exists (Boserup 1965), or when there is a recognized possibility for an individual or group to increase production and thereby secure an economic advantage (Netting 1977, 1993), will a technological innovation be adopted. Boserup (1965) has emphasized the primacy of population growth and the consequent pressure on resources. She notes that the simple knowledge or availability of techniques does not insure their use if the effect of adoption is to increase per capita work. The slow development of iron technology in the first millennium eastern Mediterranean by the major civilizations of the area (long after the knowledge of the technology was widespread) seems to be an archaeological example of this phenomenon (Forbes 1955-1972; Wertime and Muhly 1980).

A second area where we could question Childe is whether technological innovation necessarily leads to the creation of a surplus. Ethnographic and historical evidence suggest that no surplus will be created in the absence of systemic pressure to intensify production through technological innovation or without some possibility for a perceived benefit to an individual or group. A technological innovation may allow an increase in energetic efficiency and yet be adopted for the purpose of reducing energetic input (time and effort) rather than of increasing production and output. In other words, without pressure, either population pressure or the advantage gained by a specific population segment (e.g. the elite) by increasing productivity, a technological development may be used simply to reduce the amount of work necessary to maintain production at the same level rather than to increase production above existing levels of need. The adoption of guns and horses by the Plains Native Americans, in the absence of any greater production, is one example.

A third question that can be raised is whether surpluses are always used to the economic and political advantage of an individual or group. Again, the answer is no. In many societies where surpluses beyond subsistence needs are produced, leveling mechanisms channel surplus wealth into the reinforcement or validation of existing social relations where they are consumed. An example would be Trobriand yam production, where there are plenty of incentives to produce a surplus which in good years rots and is unused by the chief (see Weiner 1976). There are in these societies few or no opportunities to invest the surplus in the expansion of personal or group control of resources or political power. This is not to say that surpluses are not used to the advantage of an incipient or developing elite; only that the simple existence of a surplus does not automatically stimulate the investment of the surplus in an attempt to consolidate power. The motivation must come from elsewhere: the creation of a surplus (through technological innovation of otherwise) does not necessarily imply its use as a political or economic tool.

The response to the other key component of Childe's argument (the role of technology in the development of economic specialization) is much the same. While not denying that there is a relationship between the two, his deterministic aspect must be rejected. Childe's argument is that the creation of a surplus freed a major portion of the population from subsistence tasks and thus made specialization possible. The encouragement given to technological development by the possibilities inherent in a (technologically created) surplus further aided specialization by expanding the mass of specialized technical knowledge. As the body of technical knowledge grew and the time and effort necessary to master a technical skill increased beyond the capabilities of part-time specialists, full-time specialization was encouraged. With the growth of full-time specialization, the necessity arose for the transmission of an enlarged body of technical knowledge from generation to generation through institutionalized means of technical education. Although the dependence of society on a developing technology increased, full-time specialization widened the distance between technological specialists and the rest of society. The requirements of training and apprenticeship and the physical isolation demanded for many technical processes (e.g. smelting) were among the factors which tended to increase both the solidarity of full-time specialists and their social distance from others (see, for example, Forbes 1955-1972: VII, 228-237; Burford 1972).

The problem with this argument is that increased specialization is simply one response to the possibilities created by the availability of a surplus. Freed from the demands of subsistence, it is not obvious why members of a society should choose to become full-time specialists rather than, for instance, simply maintain their former work levels and spend a greater amount of time on leisure activities (see, for example, the hunter-gatherer data presented in Lee and DeVore 1968). The evolutionary possibilities inherent in technological specialization would not be apparent. Once again, it may be suggested that only pressure on resources, perceived as a threat to an existing way of life, could motivate individuals to invest the greater amount of energy in work required by full-time specialization.

Childe argued as well that the creation of a surplus by technological means made possible more leisure time in which to elaborate the arts and sciences characteristic of civilization. This argument is also easily rebutted. Although a certain amount of leisure time is undoubtedly necessary if non-subsistence activities are to be pursued, the existence of leisure time is not sufficient to explain the development of the arts and sciences. In the absence of some economic or social incentive for the development of this aspect of culture, it is not obvious why free time should be spent in artistic, intellectual, or scientific pursuits rather than simply in rest or play (Raber 1984).

Technology and Energetic Transformations

White (1943, 1949a, 1949b, 1957, 1959) also treated technological innovation as the basis of cultural evolution: 'the type of social organization, art, and philosophy of a given cultural system will be determined in form and in content by the underlying technology' (1949b: 378). In these views White simply echoes the view of Marx that '[i]n acquiring new productive forces men change their mode of production, and in changing their mode of production, their way of earning their living, they change all their social relations. The hand-mill gives you society with the feudal lord, the steam-mill society with the industrial capitalist' (Marx 1963:119).

Technological innovations were, in White's view, responsible for the increases in energetic efficiency characteristic of cultural evolution, as expressed in his dictum: 'culture evolves as the amount of energy harnessed per capita per year is increased or as the efficiency of the instrumental means of putting the energy to work is increased' (1949a: 368-369). The details of the mechanism of evolutionary change are omitted and, according to White, irrelevant (1957). White dealt with both technology and evolution at a very general level. The inadequacies of White's formulation have been pointed out by Carneiro (1971, 1974). Carneiro notes that cultures may evolve in the absence of any increase in the efficiency of energy utilization and suggests a comparison of the Aztec from the Basin of Mexico and the Kuikuru from the Amazon Basin as an example of the evolution of a complex society without a major advance in energetic efficiency or technological level. However, White's assertion that cultures evolve as the amount of energy harnessed per capita increases seems to hold true. The main value of White's work was in focusing attention on the crucial role of energy and energetic transformations to technology and innovation. While both White and Childe may be criticized, as Carneiro (1974) has very effectively done, they pointed out that the significance of technology in evolution lays in its role in energetic transformations and use.

A more comprehensive view of the role of technology in effecting energetic transformations and their part in cultural evolution has been developed by Richard N. Adams (1975). Adams considers the importance of energetic transformations and their control as the basis of social power. He argues convincingly that the control of energetic transformations (through the control of technology, resource exploitation, labor, etc.) can allow the concentration of power in the hands of an individual or group who understand and use the potential inherent in this control. Adams places the question of technology and cultural evolution in the context of adaptation and selection and thus develops one of the few coherent evolutionary analyses of the subject.

Julian Steward (1949, 1951, 1955) attempted to define the role of technology in a multilineal evolutionary perspective. Technology formed a fundamental part of the 'culture core': 'the constellation of features which are most closely related to subsistence activities and economic arrangements', and 'which empirical analysis shows to be most closely involved in the utilization of the environment in culturally prescribed ways' (Steward 1955:37). As such, technology was a basic concern in a cultural ecological analysis of the ways in which societies adapted themselves to their environments. Differences in cultural evolutionary paths could be explained in terms of environmental variables and technological adaptations. However, Steward recognized that the evolutionary significance of technology varied greatly according to existing social arrangements and the evolutionary potential of the environment. Technology could, in Steward's view, be a significant causal factor in evolution but technologies 'may be used differently and entail different social arrangements in each environment' (1955: 38). Steward thus emphasized that the evolutionary potential of technology must be analyzed in terms of environmental variables. He further implied that technologies were related to energetic transformations and energy use, although the mechanism was left unspecified. Steward's 'levels of sociocultural integration' (1951, 1955: 43-63) were presumably, though not explicitly, related to levels of energy use and, therefore, to technology, although the emphasis was on organizational size and complexity rather than on the technological adaptation which underlay the social organization.